Tableau Desktop Cookbook
Quick & Simple Recipes to Help You Navigate Tableau Desktop

Lorna Brown

Beijing · Boston · Farnham · Sebastopol · Tokyo

Tableau Desktop Cookbook

by Lorna Brown

Published by O'Reilly Media, Inc., 1005 Gravenstein Highway North, Sebastopol, CA 95472.

O'Reilly books may be purchased for educational, business, or sales promotional use. Online editions are also available for most titles (*http://oreilly.com*). For more information, contact our corporate/institutional sales department: 800-998-9938 or *corporate@oreilly.com*.

Acquisitions Editor: Angela Rufino		**Indexer:** nSight, Inc.	
Development Editor: Michelle Smith		**Interior Designer:** David Futato	
Production Editor: Katherine Tozer		**Cover Designer:** Karen Montgomery	
Copyeditor: nSight, Inc.		**Illustrator:** Kate Dullea	
Proofreader: Sharon Wilkey			

September 2021:   First Edition

Revision History for the First Edition

2021-08-02:   First Release

See *http://oreilly.com/catalog/errata.csp?isbn=9781492090113* for release details.

978-1-492-09011-3

[MBP]

Table of Contents

Part I.   Starting at the Very Beginning

Part II.   Building a Foundation

Part III.   Broadening Your Data Viz Knowledge

Part IV. Advancing Your Techniques

Preface

About six years ago, I was asked to produce a visualization based on data from Microsoft Excel, using software called Tableau. I'll admit I fumbled around a little bit at first. When I had finished creating the visual analysis, I presented that back to stakeholders, and I saw their faces light up. For them, seeing their data in this new visualization enabled them to ask more questions and act on the data they were seeing. Since that light bulb moment, I saw the power of data visualization and wanted to empower others to work with data. But what is data?

What Is Data?

Data is all around us. Data is facts and statistics that are collected about something. That could be anything—sales, health and fitness, housing, or flights. Each one of those areas collects data. Data can be stored in many forms, from Excel spreadsheets to databases, and each organization will have a different storage process and many types of data. Without data, we can't create data visualizations. "But what is data visualization?" I hear you ask.

An Introduction to Data Visualization

Data visualization is the visual representation of data points to communicate the messages within your data set more effectively to your audience. Data visualization enables your user to see and understand their data, ask further questions that they couldn't before, and make data-informed decisions from these visualizations.

When there is visual representation of data, like a chart, we can quickly gain insight and spot trends or outliers within the data. Data visualization is more effective when preattentive attributes are used well to communicate with data, because these attributes allow us to process the data almost immediately. If you want to understand more about how to communicate with data, see Carl Allchin's *Communicating with Data* (O'Reilly, 2021, currently in Early Release).

This book will help you get the most out of following 10 preattentive attributes, which you need to be aware of before creating any data visualizations:

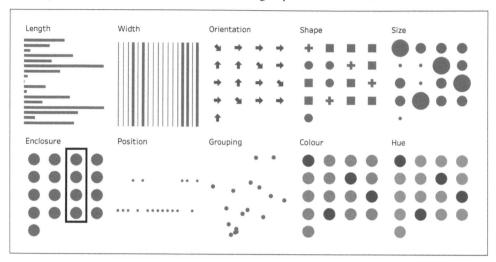

Each attribute is used because it allows our brains to process this information almost immediately, to quickly see patterns or trends easier. Some attributes are more effective than others; length, for example, is the easiest attribute to read. These attributes will be mentioned throughout this book.

An Introduction to Tableau

Tableau Software offers several products. To start with, Tableau Prep Builder helps with data preparation. Tableau Online or Tableau Server allows you to share and collaborate with colleagues on data sources and dashboards. Finally, Tableau Desktop is a data visualization tool and the focus of this book.

Why I Wrote This Book

I started teaching Tableau to myself more than six years ago now. As a student, I was used to learning through books. Throughout my journey, from student to a Tableau Zen Master, I have always thought it important to read blog posts and other books and to watch training videos.

 A Tableau Zen Master is a selective program by Tableau; there are currently 43 in the world. To become a Zen Master, you need to show a deep understanding of how Tableau works, help teach other Tableau users, and collaborate with others.

During my journey, I was taught by the best of the best within The Information Lab's Data School program, teaching the next generation of analysts. I have also enjoyed teaching other cohorts and the community. I help lead Workout Wednesday, a community initiative helping you build your Tableau skills, and I have published Tableau Tips on a regular basis. I love being able to share my knowledge with everyone. Writing this book has allowed me to share much of what I have learned and been taught over the last six years. I have condensed my knowledge into easy-to-follow recipes so you don't have to struggle, as I did, when first learning Tableau. I also add my own sprinkle of tips and tricks to make your Tableau better.

Who Is This Book For?

I recommend that less experienced or new users of Tableau read cover to cover, as I have structured the flow of chapters deliberately. If you are more experienced, you might be able to read chapters in isolation, especially for the newer features of Tableau, but you will learn more tips and tricks if you also read cover to cover.

How the Book Is Organized

This book has four parts. They have been ordered to go from basics to intermediate to advanced-level Tableau, which allows you to build on the skills throughout each chapter.

Part I: Chapters 1–2
> The first two chapters will take you through how to connect to various data sources, how you can relate them in different ways, and then how to get to grips with the Tableau workspace.

Part II: Chapters 3–7
> Once you have the knowledge of the data sets being used throughout this book and the workspace we will be using, Chapter 3–Chapter 7 will start giving you the understanding of the basic chart types in Tableau. By the end of the first seven chapters, you will know how to build your first dashboard.

Part III: Chapters 8–13
> Within these chapters, you will continue to learn about more advanced chart types and how you can format some elements within each of the charts. You will also learn how to use a story, which guides your users through a piece of analysis that you want to show them.

Part IV: Chapters 14–19
> The final few chapters will delve into advanced techniques to enhance the way you interact with your analysis. They will also teach you some of the more complex calculations, all with enhancing your analysis in mind. You will finish

off by creating more dashboards that have the interactivity from the previous chapters.

Chapter 20
    Here, you will find a collection of 20 ways to continue your Tableau Desktop learning journey, with both a social and technical approach.

Each solution should be a new or duplicated worksheet that will allow you to go back to an individual recipe.

You can find the completed workbook and data on GitHub (*https://github.com/ LornaBrownTIL/Tableau-Desktop-Cookbook*). Download the data to get started with this book.

Outcomes of the Book

By the end of this book, you will understand how to do the following:

- Build basic and complex data visualizations in Tableau Desktop
- Use the newer features, including set and parameter actions
- Improve your current analytical analysis

And then you will be able to:

- Apply the skills learned using data visualization best practices
- Create interactive dashboards to support business questions
- Have enough reference materials to take Desktop Specialist and Tableau Desktop Certified Associate exams

Conventions Used in This Book

The following typographical conventions are used in this book:

Italic
    Indicates new terms, URLs, email addresses, filenames, and file extensions.

`Constant width bold`
    Shows commands or other text that should be typed literally by the user.

This element signifies a tip or suggestion.

This element signifies a general note.

This element indicates a warning or caution.

Using Code Examples

Supplemental material (code examples, exercises, etc.) is available for download at *https://github.com/LornaBrownTIL/Tableau-Desktop-Cookbook*.

If you have a technical question or a problem using the code examples, please send email to *bookquestions@oreilly.com*.

This book is here to help you get your job done. In general, if example code is offered with this book, you may use it in your programs and documentation. You do not need to contact us for permission unless you're reproducing a significant portion of the code. For example, writing a program that uses several chunks of code from this book does not require permission. Selling or distributing examples from O'Reilly books does require permission. Answering a question by citing this book and quoting example code does not require permission. Incorporating a significant amount of example code from this book into your product's documentation does require permission.

We appreciate, but generally do not require, attribution. An attribution usually includes the title, author, publisher, and ISBN. For example: "*Tableau Desktop Cookbook* by Lorna Brown (O'Reilly). Copyright 2021 Lorna Brown, 978-1-492-09011-3."

If you feel your use of code examples falls outside fair use or the permission given above, feel free to contact us at *permissions@oreilly.com*.

O'Reilly Online Learning

O'REILLY® For more than 40 years, *O'Reilly Media* has provided technology and business training, knowledge, and insight to help companies succeed.

Our unique network of experts and innovators share their knowledge and expertise through books, articles, and our online learning platform. O'Reilly's online learning platform gives you on-demand access to live training courses, in-depth learning paths, interactive coding environments, and a vast collection of text and video from O'Reilly and 200+ other publishers. For more information, visit *http://oreilly.com*.

How to Contact Us

Please address comments and questions concerning this book to the publisher:

O'Reilly Media, Inc.
1005 Gravenstein Highway North
Sebastopol, CA 95472
800-998-9938 (in the United States or Canada)
707-829-0515 (international or local)
707-829-0104 (fax)

We have a web page for this book, where we list errata, examples, and any additional information. You can access this page at *https://oreil.ly/tableau-desktop-cookbook*.

Email *bookquestions@oreilly.com* to comment or ask technical questions about this book.

For news and information about our books and courses, visit http://oreilly.com.

Find us on Facebook: *http://facebook.com/oreilly*

Follow us on Twitter: *http://twitter.com/oreillymedia*

Watch us on YouTube: *http://youtube.com/oreillymedia*

Acknowledgments

Many people in my life have influenced and supported me throughout this journey. Firstly, the person who puts up with the most is my wonderful husband, Michael Brown. He has been there through the sad and happy times, and his continuous love and support in everything that I do is much appreciated. Thank you for keeping me sane throughout lockdown; now that we have survived our first year of marriage in lockdown, I'm sure there will be plenty more normal years to come! I must not forget to mention our crazy cats, Garfield and Gizmo—thank you for keeping me warm and stress-free with your constant need for attention and your craziness. And although I have not been able to see them much over the past year, my parents, grandparents, and family have supported me through everything.

I started my Tableau journey with the Information Lab's Data School in 2015. Without it, I would not be where I am today—big thanks to Tom, Robin, and Craig for everything you do for the company, and thank you for taking a chance on me back in 2015. With two special mentions—first, thank you to Andy Kriebel for teaching me mostly everything about Tableau and constantly pushing me to achieve my best; and second, thank you, Carl Allchin, for being there throughout my many rants and frustrations over the years, and becoming someone I know I can turn to no matter what. Appreciation also to Stephanie Kearns, for being my partner in crime over the past year.

Once I started learning Tableau, I joined the fantastic Tableau Community, which is one of a kind. The community has been a wealth of knowledge, inspiration, and friendships. I have learned so much from the community, and now it is my time to give back. Thank you to everyone within the Tableau Community who has helped me on my journey so far; you know who you are.

A special mention goes out to Workout Wednesday leaders, past and present. It has been a pleasure to work alongside all of you. Thank you to all the Tableau Zen Master and Ambassadors for everything you do to make the community even better, and to the past and present Tableau employees who have helped the community grow.

To the reviewers of this book, thank you for keeping me on the straight and narrow, and refraining me from many tangents to keep the content as simple as possible. Also, a special thanks to Angela, my editor, and O'Reilly for giving me this opportunity to help teach people Tableau.

Starting at the Very Beginning

CHAPTER 1

Introduction to Data

Before you start to use Tableau, you will need data. Data comes in many different shapes and sizes, from Microsoft Excel files to databases. Data is all around us, and more and more companies need the ability to see and understand their data, which is Tableau's focus. Tableau Desktop has a variety of data connection options, from files to databases to saved data connections. This chapter will explore how to connect and use different data sets within your Tableau workbook.

1.1 Connecting to Data

When you first open Tableau, you will see the Start Page. The Start Page consists of three panes: Connect, Open (and recently connected workbooks), and Discover.

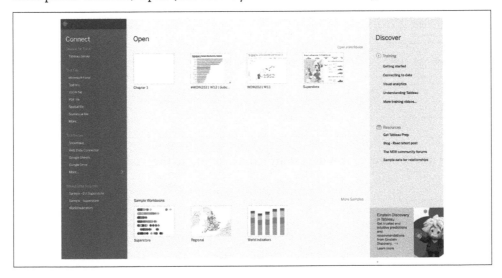

Within the Connect pane, you can see four sections: Search for Data, To a File, To a Server, and Saved Data Sources.

1.2 Searching for Data

The Search for Data option allows you to connect to a data source that has been published onto your Tableau Server or Tableau Online platform.

Problem

You want to connect to a data source on your Tableau Server/Online instance.

Solution

1. On the Start Page, click Tableau Server:

2. This will ask you to log in to your Tableau Server instance. (For more information, refer to your Tableau Server Administrator.)

3. Once you have connected to your Tableau Server instance, you will see this pop-up box that lists all the data sources available on your server. You can select one and click Connect:

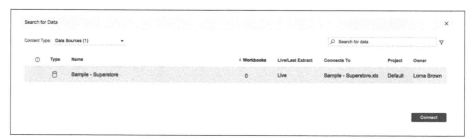

4. This will connect to that data source and take you to the Tableau Workspace:

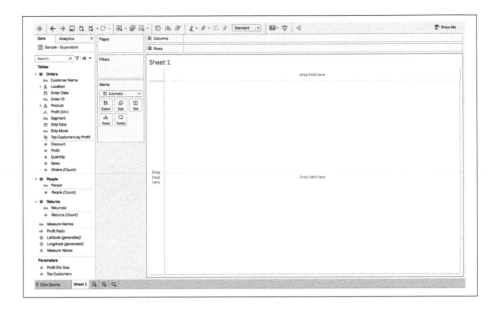

Discussion

Your company might have a set of trusted data sources that you should be using to do your analysis, which is why you would want to connect to the Tableau Server data source. I could write a whole book on how to set up your Tableau Server data sources, but this book is focused on Tableau Desktop. If you want to prepare your data sources using Tableau Prep, Carl Allchin's *Tableau Prep* (O'Reilly) will get you started.

If you want to know more about Tableau Server, ask your IT department or your Tableau Server Administrator.

1.3 Connecting to a File

Connecting to a file is probably the most common way that people connect to data, especially when they are first starting to use and understand Tableau.

Problem

You want to connect to an Excel file stored on your computer.

Solution

1. On the Start Page, under the To a File section, click Microsoft Excel:

2. This will bring up the File Explorer on Microsoft Windows (Finder on Mac) to allow you to search for the Excel file:

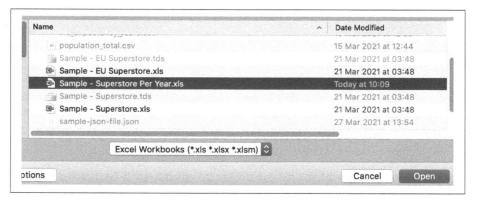

3. Once you have clicked Open, Tableau will take you to the next page, the data source page:

4. This Excel file has several sheets, which are different tabs in the file. We want to use the Orders 2021 sheet:

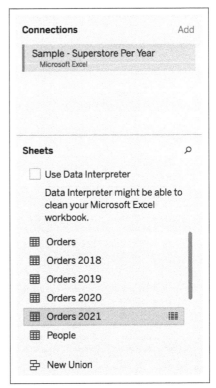

5. Double-click the Orders 2021 sheet to automatically add it to the canvas within the data source page:

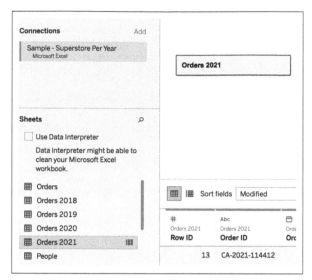

You are now ready to start your analysis on 2021 orders:

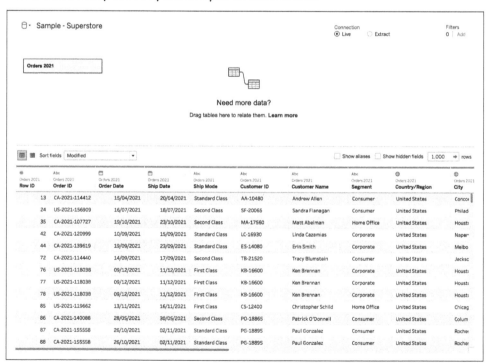

Discussion

Connecting to a file is the easiest way to get started with Tableau Desktop because it still gives you several different file types. We connected to a Microsoft Excel file, but you can also connect to other files: Text, JSON, PDF, Spatial, Statistical, and More. When you select a specific type, the search box that pops up will allow you to click only that specific type of file; therefore, if you are unsure of the file (for example, whether it's a CSV or Excel), use the More option instead. Also, Tableau has its own file type called Hyper files, and you can connect to these through the More option as well.

We double-click a sheet to add it to the view. You can also drag and drop to the canvas section in the middle:

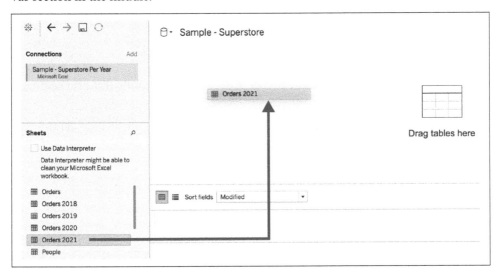

When you've brought your data onto the canvas, you will see the data grid at the bottom. This allows you to preview a sample of the data before you start your analysis:

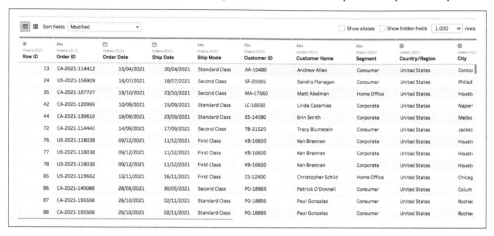

When connecting to the different file types, there might be slightly different settings.

A CSV (comma-separated values) connection will automatically bring the CSV onto the data source canvas once you click Open, and the left pane will show all the CSV files in the same location as the one you have opened:

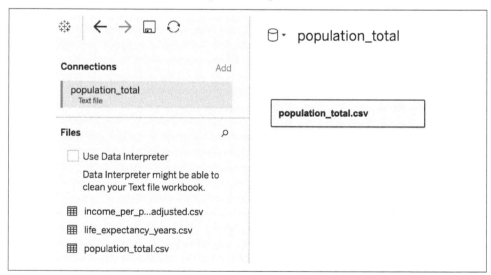

When connecting to a JSON file, Tableau will ask you to select the schema you want to bring into the view, which allows you to bring in certain levels of detail from your JSON files:

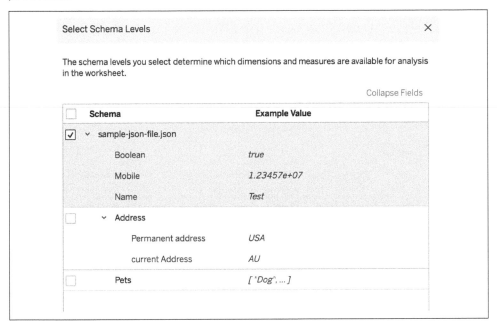

When connecting to a PDF file, Tableau will ask you which pages you would like to look at. The PDF input will look for tables within the structure of the PDF:

If you need to connect to a spatial file or a statistical file, once you have connected, Tableau will automatically bring in the file to the data canvas (like when connecting to a CSV), unless there are multiple tabs/sheets, in which case you will have to add the sheet to the canvas manually.

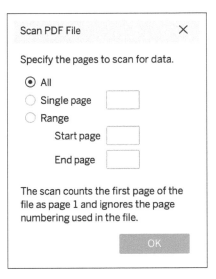

1.4 Messy Data Sources

When you are new to working with data, you will appreciate a clean and tidy data source, but sometimes you might be given a messy data source or one that needs cleaning. This could range from simple data preparation to something more complex.

Problem

You want to connect to a formatted Excel spreadsheet that is nicely styled for Excel use, where the actual headers do not start in the first row:

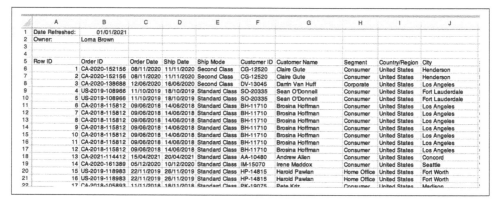

Solution

1. Connect to the Excel spreadsheet, and when you add the Orders sheet to the canvas, you will notice that in the data grid the data hasn't been brought in correctly:

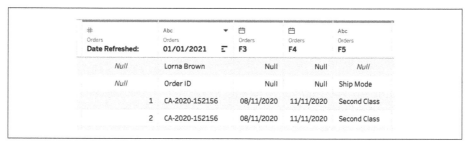

2. To rectify this, in the left pane under Sheets, find the Use Data Interpreter checkbox:

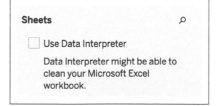

3. Check this box to have the Data Interpreter look at your file and try to understand what data you actually want to use. Tableau then removes the unnecessary
rows that don't match the other rows:

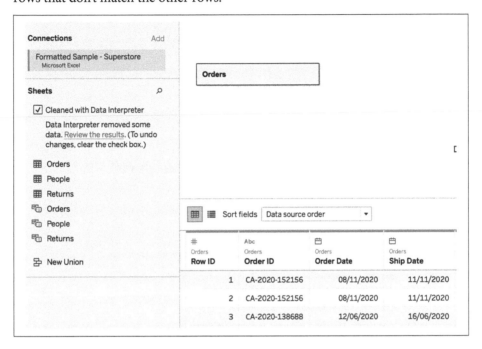

Discussion

When you are using the Data Interpreter, Tableau also gives you the option to review
the results:

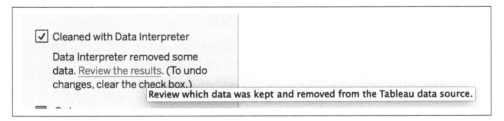

When you select "Review the results," Tableau will automatically open an Excel spreadsheet to show you the changes made:

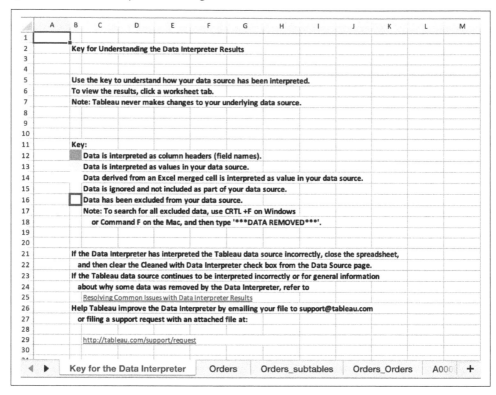

Using the Data Interpreter will not change your raw underlying data; it will change only the data that is being viewed and used in Tableau.

Pivoting Data

When working with Microsoft Excel, CSVs, Google Sheets, and PDFs, you have the ability to pivot data. You might want to pivot data because Tableau prefers a specific structure. For example, if you have multiple date columns, being able to analyze the data will be difficult; therefore, you can pivot the data.

Here is an example:

Sub-Category	January 2018	February 2018	March 2018	April 2018	May 2018	June 2018	July 2018	August 2018
Accessories	827.89	1,120.99	1,108.93	2,545.99	695.49	667.04	3,308.69	1,937.73
Appliances	312.58	89.92	502.96	532.58	918.60	2,275.25	81.74	2,096.75
Art	176.99	73.66	413.27	567.92	288.22	686.11	256.04	203.66
Binders	814.51	339.26	1,525.68	985.75	4,372.30	4,275.79	2,934.49	4,251.39
Bookcases	1,010.06	Null	1,706.45	308.50	640.63	759.57	1,487.67	794.28

Tableau would prefer that all the dates be in one column. Therefore, we need to pivot the data, which will increase the number of rows but reduce the columns to three. To do this, click the first date column, press Shift, and scroll to the right to select the last date column, then right-click and select Pivot:

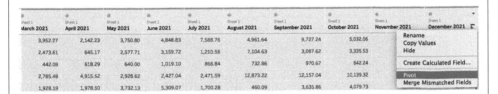

March 2021	April 2021	May 2021	June 2021	July 2021	August 2021	September 2021	October 2021	November 2021	December 2021
3,952.27	2,142.23	3,750.80	4,848.83	7,588.76	4,961.64	9,727.24	5,032.06		
2,473.61	645.17	2,577.71	3,159.72	1,210.56	7,104.63	3,087.62	3,335.53		
442.08	618.29	640.00	1,019.10	866.84	732.86	970.67	642.24		
2,785.49	4,915.52	2,928.62	2,427.04	2,471.59	12,873.22	12,157.04	10,139.32		
1,928.19	1,978.50	3,732.13	5,309.07	1,700.28	460.09	3,635.86	4,079.73		

The pivot will give you two new columns, Pivot Field Names and Pivot Field Values. And your new data structure will look like this:

Pivot Field Names	Pivot Field Values	Sub-Category
April 2018	2,545.99	Accessories
April 2019	1,647.03	Accessories
April 2020	1,616.46	Accessories
April 2021	2,142.23	Accessories
August 2018	1,937.73	Accessories
August 2019	1,372.45	Accessories
August 2020	4,104.32	Accessories
August 2021	4,961.64	Accessories

You can rename these fields by double-clicking the name to edit:

Pivoting data can also be useful when you have the same measures across different columns.

1.5 Connecting to a Server

Connecting to a server is different from searching for data through Tableau Server. This type of connection allows Tableau to connect to databases or cloud-based data, like Google Sheets.

Problem

You want to connect to a Google Sheet.

Solution

1. In the Connect pane, under To a Server, select More, and then select Google Sheets:

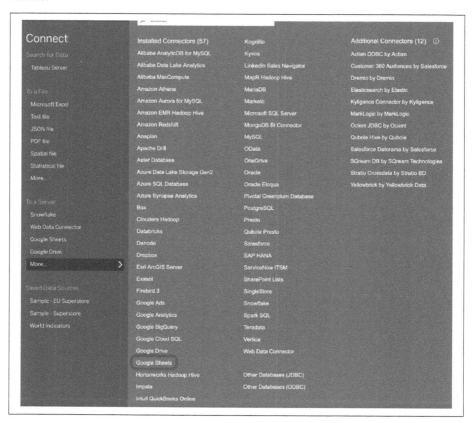

2. Tableau will open a browser window and ask you to log in to your Google account; log in.

3. Google will ask you to give authorization to allow Tableau to see and download all of your Google Drive files:

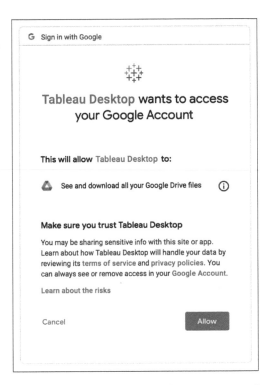

4. Once you have trusted Tableau, you are able to see the Google Sheets associated with your account. You can use any data source and click Connect:

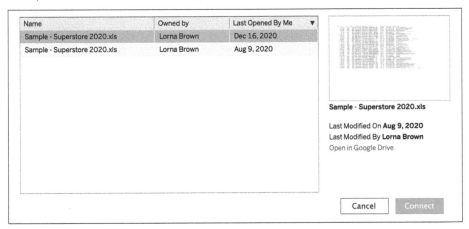

Discussion

To a Server allows you to connect to different database types and cloud-based data. If you want to connect to a specific server, you will need to provide connection details to the Server IP or connection string and a username and password:

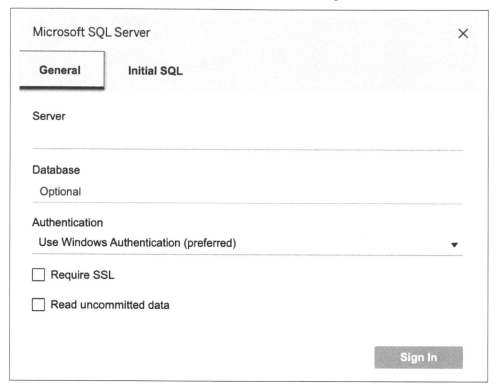

If you are connecting to a database for the first time, you might need to download the associated driver. If you do need to install a driver, Tableau will tell you where to go to get the driver.

1.6 Saved Data Sources

The final section on the Connect pane on the Start Page is for connecting to Saved Data Sources.

Problem

You want to connect to a saved data source, and you want to add a new saved data source for instant reusable access.

Solution

1. In the Connect pane, select one of the Saved Data Sources. This will take you straight to a new worksheet ready to start your analysis.

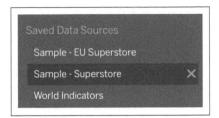

2. To add a data source to your Saved Data Sources, right-click Data at the top left of the Tableau interface and choose "Add to Saved Data Sources" (see Chapter 2 for more information).

3. Tableau will then ask you to save this file into your Tableau Repository as a .tds file, which is a Tableau Data Source file. Save this connection, and the file will appear under the Saved Data Sources section on the Connect pane.

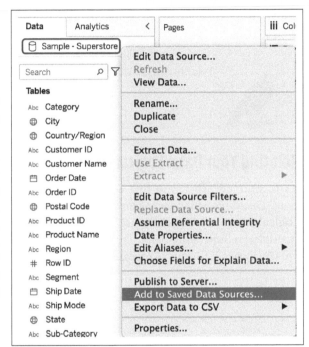

Discussion

Saved Data Sources gives you quick access to data sources that you regularly use. You can also repeat the process for any database connection to save the connection string.

1.7 Connecting to Secondary Data Sets: The Data Model

A *data model* is like a diagram telling Tableau how to query your data. The data model consists of two layers: the *logical layer* and the *physical layer*. When you first connect to any data within Tableau, you start in the logical layer. If you want to add additional data, you will have to decide between the logical or physical layer.

The *logical layer* uses Tableau's data model to create relationships between two or more data tables based on one or more fields in each data table. Relationships are dynamic and flexible, as they query the data only when fields are being used from that table. Relationships also allow your data to be at different levels of detail, using many-to-many situations.

In the *physical layer*, you can create joins and/or unions with data. The physical layer is not dynamic and flexible like the logical layer and is best suited for data that is at the same level of detail, therefore creating a one-to-one join type. If you use the physical layer for data at different levels of detail, you could cause duplication in your data. See "Joining Your Data: The Physical Layer" on page 32 for more information about joining data.

Prior to Tableau 2020.2, Tableau had only the physical layer.

Relating Your Data: Logical Layer

As mentioned, when you first connect to data, you are automatically in the logical layer. Deciding which layer you need will depend on the type of data you are using. Logical layers are especially useful when your data has varying levels of detail. For example, in Superstore Sales, the Orders table has one row per order per product, so the Customer ID/Name is repeated over several rows:

# Orders 2021 **Row ID**	Abc Orders 2021 **Order ID**	🗓 Orders 2021 **Order Date**	Abc Orders 2021 **Customer ID**	Abc Orders 2021 **Product ID**	Abc Orders 2021 **Category**	Abc Orders 2021 **Sub-Category**	Abc Orders 2021 **Product Name**
76	US-2021-118038	09/12/2021	KB-16600	OFF-BI-10004182	Office Supplies	Binders	Economy Binders
77	US-2021-118038	09/12/2021	KB-16600	FUR-FU-10000260	Furniture	Furnishings	6" Cubicle Wall Clock…
78	US-2021-118038	09/12/2021	KB-16600	OFF-ST-10000615	Office Supplies	Storage	SimpliFile Personal F…

You may now want to add customer details data to find out more about your customers. This relationship will not be one-to-one because Customer ID appears more than once in the Orders data set. This is where we can use relationships.

Problem

You want to relate your orders data to your customer details data to allow analysis on the customer details.

Solution

1. We first need to add this additional data to our Data Source page. To do this, click Add (next to Connections) and select Microsoft Excel:

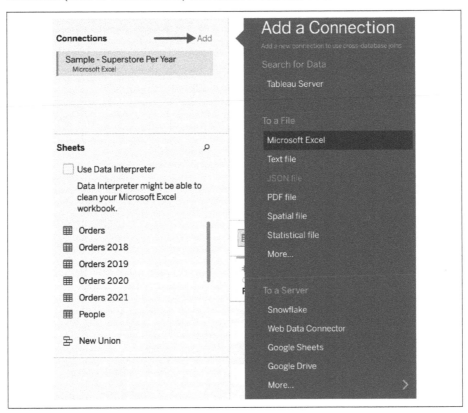

2. On selection of the file type, you will be asked to find your file:

In the left pane, you will now have two data sources connected. Now we have both data sources, and we need to create the relationship.

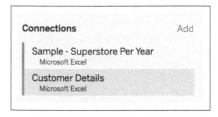

3. To do this, drag and drop the Customer Details sheet onto the data source canvas, which creates an orange connection, known as a *noodle*:

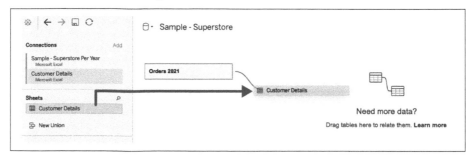

Tableau now gives you a pop-up box to ask you how you want to relate your data:

4. In this case, Tableau has automatically picked up the field that is common from both data sets. But if this option is empty because the field names are named differently, you can tell Tableau which fields you want to relate your data to:

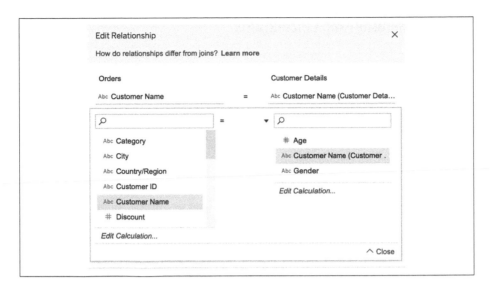

5. Finally, if you have more than one field you want to relate to, you can select the "Add more fields" option to define the second relationship:

Discussion

A relationship allows you to relate your data at different levels of detail and is the default way to connect multiple tables. For example, you have data at an individual orders level, but you have sales target data on a monthly basis. With relationships, you can create a relationship calculation.

Creating Relationship Calculations

Problem

You want to create a relationship between your Daily Orders table and your Monthly Targets data by using a relationship calculation, because your Orders data doesn't contain a month of order field.

Solution

1. Add a new data connection to the Monthly Targets data:

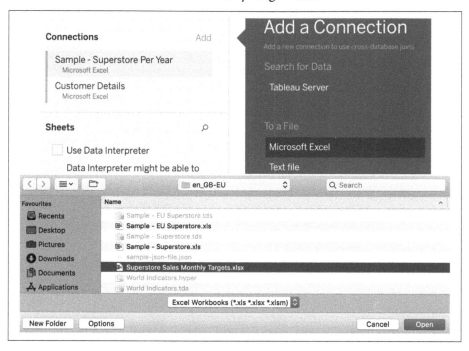

2. Drag the Monthly Targets data to the data source canvas:

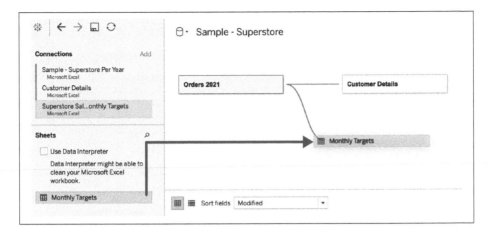

3. This will pop up for you to define the relationship between the data sets. Tableau has automatically found that Sub-Category is a common field. But to add a relationship calculation, select "Add more fields":

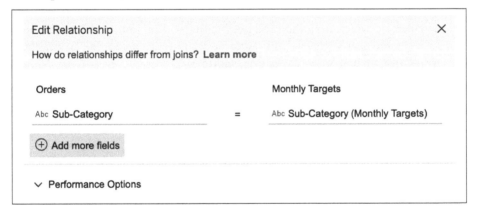

4. This will populate the drop-down of the current fields in the data. At the bottom, select Create Relationship Calculation.

Once you have selected the Create Relationship calculation, you should use this calculation, which uses Tableau's DATETRUNC function to aggregate the Order Date to the specific level. This could be year, quarter, month, or week/day (for more information about dates, see Chapter 5).

Relationship Calculation

```
DATE(DATETRUNC('month',[Order Date]))
```

5. Now you can select the Month field from the Monthly Targets data:

Discussion

The reason you want to create a relationship calculation could be that your original data doesn't contain the correct field that you need to relate to. If you don't have this option, you need to make sure you are in Tableau version 2020.3 or above.

Now that you have your targets data included, you'll notice that the targets go back to 2018, but we have brought in only our 2021 data.

Unioning Your Data: The Physical Layer

Our Excel file has multiple tabs for each year of our data; this is a common Excel approach. The tabs all have the same structure of data, but they contain different years. We need the ability to stack the Excel tabs on top of each other to get the full date range in our data set. To do that, we need to use a technique called a *union*. A union stacks data on top of each other when the data has a common set of fields. This happens within the physical layer of the data model.

Problem

You want to bring several tabs of orders data within Excel together to create one large orders data set.

Solution

1. To get to the physical layer editor, right-click the Orders 2021 logical layer. and select Open, or double-click the primary table:

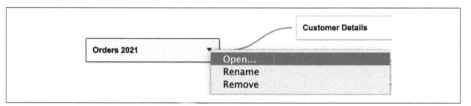

2. From here, you can drag and drop the specific sheets underneath the 2021 sheet, which will create a union. As the box suggests, drag the table to union the data:

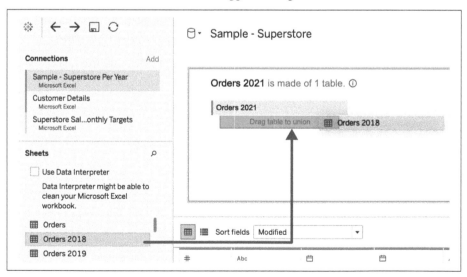

3. Repeat for the sheets Orders 2019 and Orders 2020.

4. If you find you have added too many sheets, you can edit the union by right-clicking Orders 2021+ and selecting Edit Union:

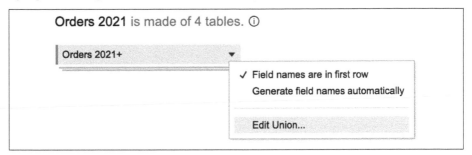

This shows you which sheets are now stacked on top of each other:

Discussion

A union is bringing similarly structured data together by stacking different files or tabs on top of each other.

You can also create a *wildcard union*. A wildcard union follows a naming convention you suggest, and then it will bring those sheets together. When you go to edit a union, you will see an option at the top for "Wildcard (automatic)":

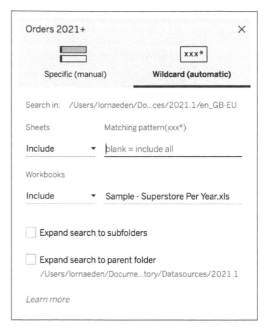

Within the settings, you can tell Tableau how you want to define the wildcard with a specific pattern. All of our sheets in this example start with "Orders" followed by the year:

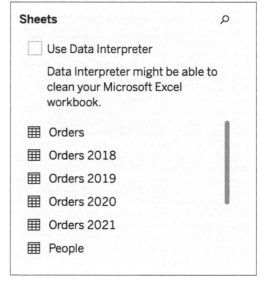

Therefore, in the matching pattern we can use **"Orders *"**. The asterisk is a symbol for any character.

This will then bring all the sheets together and is useful if you update the Excel file with a new year. Tableau will automatically bring that sheet as well, whereas you would have to manually add the new year's sheet to the specific union.

There are two ways you can see you have created a union. First, the tab you dropped onto the canvas has changed to look stacked. Second, by scrolling to the farthest right columns in the data preview reveals two new fields called Sheet and Table Name, to show you which rows of data are from which table.

Once you are happy with your union, click the close icon to close the physical layer:

Finally, you can rename any physical or logical layer by right-clicking a block and selecting Rename:

Unioning is great when you have multiple data sources that have the same common fields. But what if you want to add additional fields to your data that you know won't duplicate your data?

Joining Your Data: The Physical Layer

In addition to unioning, you can also join your data inside the physical layer. *Joining* data gives you the ability to add additional columns to your data set. With the data model, you might find that you don't need to join your data, especially when you have data that isn't a unique record, which is known as a *many-to-many relationship*. A join is still useful if you have a one-to-many or a one-to-one relationship.

Problem

You want to join your orders data, by region, to include the area manager for each region.

Solution

1. Open the Orders logical layer to get to the physical layer:

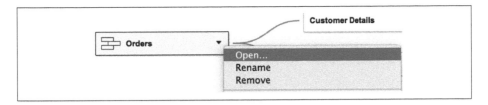

2. Double-click the People table; this will automatically create a join:

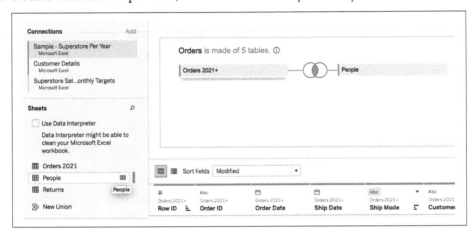

3. This has created a join based on a common field in both data sets. To find out which one, click the Venn diagram in the middle of the two tables:

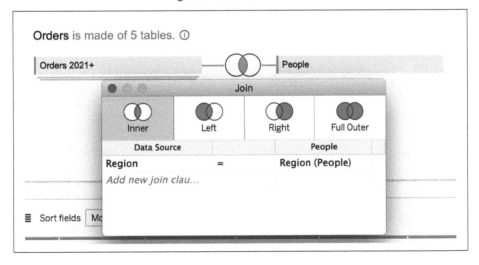

Using the default inner join type could result in mismatched rows being excluded. See the Discussion for more information about join types.

4. You can add new join clauses, depending on the level of detail of your data. To find these new columns, scroll to the right in the data grid:

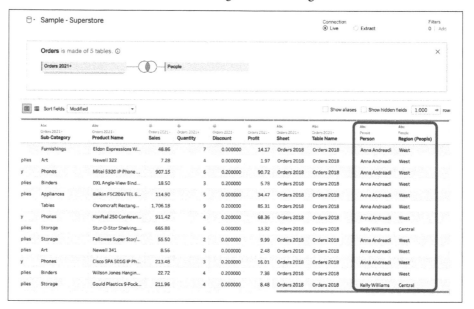

Discussion

The Venn diagram denotes the join type that you are using. Tableau Desktop has four types: Inner, Left, Right, and Full Outer.

We have two tables. Our left table contains our Orders and Region:

Order ID	Region
CA-2018-100006	East
CA-2018-100090	West
CA-2018-100328	East
CA-2018-100363	West
CA-2018-100391	East
CA-2018-100678	Central

Our right table contains Region and Person:

Region	Person
Central	Kelly Williams
East	Chuck Magee
South	Cassandra Brandow

Choose the Venn diagram that is shaded in the middle for an Inner join:

In an Inner join, we take the left and the right table, and will get a match only on the regions that are in both sets of tables—the East and Central regions. All of our orders that contain the West region will be removed from our data set, and the person for the South region won't appear either.

Order ID	Region	Person
CA-2018-100006	East	Chuck Magee
CA-2018-100090	West	Anna Andreadi
CA-2018-100328	East	Chuck Magee
CA-2018-100363	West	Anna Andreadi
CA-2018-100391	East	Chuck Magee

Choose the Venn diagram with the left circle shaded shaded for a Left join:

A Left join is going to bring back all of the data from our left table, including the records that do and don't match. So, it will bring back all of our orders, and the person would be Null:

Order ID	Region	Person
CA-2018-100006	East	Chuck Magee
CA-2018-100090	West	Anna Andreadi
CA-2018-100328	East	Chuck Magee
CA-2018-100363	West	Anna Andreadi
CA-2018-100391	East	Chuck Magee
CA-2018-100678	Central	Null

Select the Venn diagram with the right circle shaded for a Right join:

This will bring back everything from the right table that does and doesn't match by region. In this example, we will have Central, East, and South regions, but we won't have any orders for the South region:

Order ID	Region	Person
CA-2018-100006	East	Chuck Magee
CA-2018-100328	East	Chuck Magee
CA-2018-100391	East	Chuck Magee
CA-2018-100678	Central	Kelly Williams
Null	South	Cassandra Brandow

And finally, select the last join type, Full Outer (represented by both circles being shaded):

Full Outer

This join type will bring back both sets of data, regardless of whether the records join together or not; you will just have Nulls where the data doesn't match:

Order ID	Region	Person
CA-2018-100006	East	Chuck Magee
CA-2018-100090	West	Null
CA-2018-100328	East	Chuck Magee
CA-2018-100363	West	Null
CA-2018-100391	East	Chuck Magee
CA-2018-100678	Central	Kelly Williams
Null	South	Cassandra Brandow

1.8 Data Types

When you bring your data into Tableau, Tableau tries to recognize its type. For example, numbers are brought through as either integers or floats, which are decimals. If you have a field that looks like a date, Tableau will automatically use its date field, likewise with a geographic field called country or city.

Problem

You want to check that Tableau has used the correct data type on a field.

Solution

1. In the data grid at the bottom of the data source page, select the metadata grid icon.

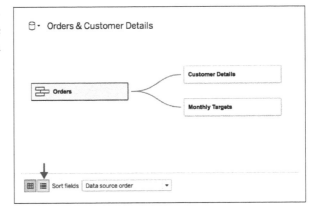

 The metadata grid shows all of your columns of data as rows. This is to enable a quick inspection of the data structure and data types.

2. You will see icons next to each field. Click an icon to check the type of data:

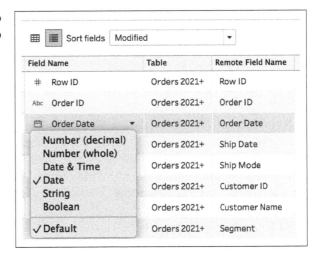

3. If you want to specifically change a geographic field, you can find the field with the globe icon and change the role:

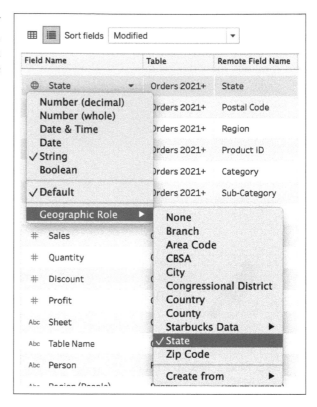

Discussion

It is important to check the data types Tableau has assigned to your data. Tableau has six types of data. Number (decimal) and Number (whole) return numeric fields as either a decimal or a whole number. Date & Time and Date both use Tableau's built-in date hierarchy (more in Chapter 5). String is used for any field that contains text. And finally, Boolean returns either true or false values within the data.

1.9 Creating Extracts

When you are connecting to big data sets, you might have performance issues caused by slower loading times of your dashboards. Creating extracts can help ease those performance issues, along with giving you the ability to use an extract offline compared to an online database.

Problem

You want to create an extract of the data source to increase performance and to use data offline.

Solution

1. Once you have connected to any data source, in the top-right corner you have the option to select Extract; click that option:

2. Now you have some options to edit the extract. Click Edit:

This will open a pop-up box asking how you want to store the data, how much data to extract, whether to have any aggregation, and how many rows:

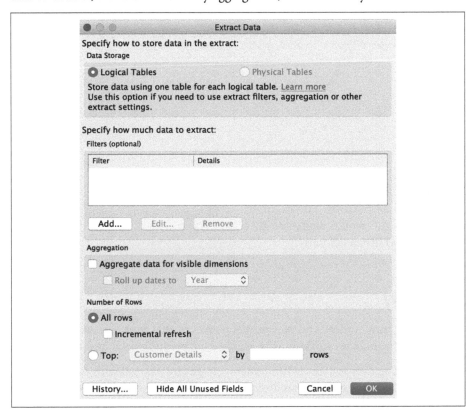

3. Fill out the information and click OK. You will need to go to a new sheet to start the extract process.

Discussion

An extract can improve performance on large data sources, but could take some time to create. An extract creates a Hyper file of your data.

You can change your extract at any time. You can create the first extract to enable faster development times, but once you have finished your dashboards productions, recommend going back to the extract to optimize. For example, I can add a filter to show only a specific subcategory of data:

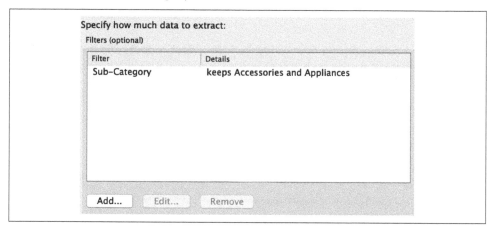

You can also aggregate the data, depending on the level of detail you use in your dashboards. For example, this is how you would aggregate to a monthly level instead of daily:

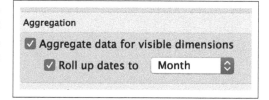

Summary

You should now understand several data sources that you can use within Tableau Desktop. This includes working with messy data, pivoting data, and creating extracts.

Understanding data sources is fundamental to getting started with data. Without data sources, you cannot start any analysis. This chapter has given you the data source you will be using for the majority of the examples in the book.

CHAPTER 2

Getting Started with Tableau

In Chapter 1, we covered how you connect to different types of data. So you have your data in Tableau, but now what? This chapter covers the Tableau Workspace, which will allow you to get a deeper understanding of how Tableau fundamentally works. Without this base-level knowledge, you might find yourself dragging and dropping data fields until Tableau does what you want.

We left Chapter 1 on the data source page, but how do you start the analysis?

2.1 Creating a New Worksheet

To get started with analyzing your data, we need to create our first worksheet.

Problem

You want to create a new worksheet to start the analysis.

Solution

1. Once you are happy with your data source, at the bottom of the data grid you will see a Go to Worksheet notification. Click Sheet 1, which is highlighted in orange:

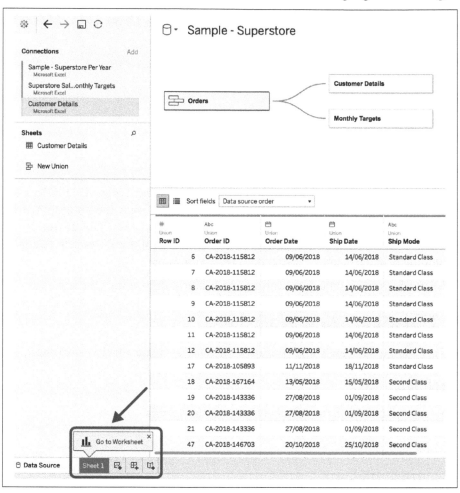

This will take you to the worksheet Tableau Workspace:

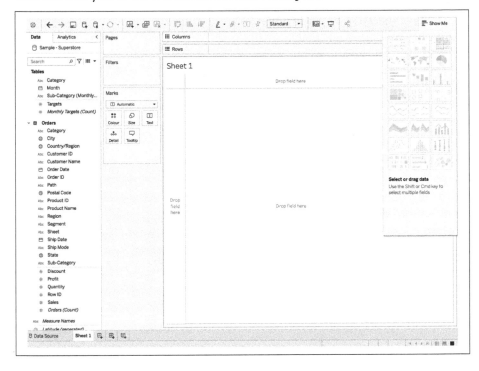

Discussion

By clicking Sheet 1, you get to a blank worksheet to start your analysis with Tableau. A worksheet is where we are going to be building all of our charts; other terminology you might see is *dashboard* (see Chapter 7) and *story* (see Chapter 13).

Once you have your first sheet, you can start building charts. You can have more than one sheet within a workbook; to add another sheet, click the first button next to Sheet 1 at the bottom:

When you are creating your views, you might want to duplicate a worksheet to be able to dig further into the exploration. If you right-click the sheet name at the bottom, you have the option to duplicate:

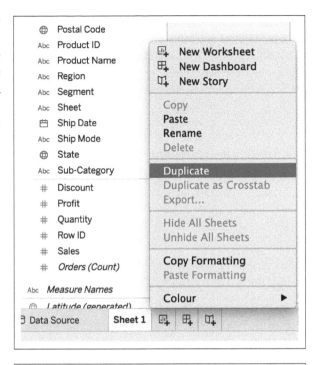

This will duplicate the worksheet and add a number to the end:

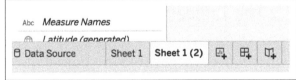

I highly recommend renaming your worksheets as you start building your charts. This is helpful not only when you come back to your workbook, but also for dashboards and the use of your workbook by someone else. To rename a worksheet, you can double-click the worksheet tab:

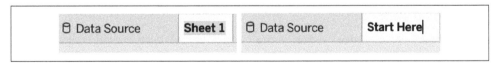

Another handy technique to help group your worksheets together is applying a color to a tab. I have seen this used in many ways, either grouped by sheets on a particular dashboard or to indicate an unfinished/finished worksheet. To change the color, right-click the sheet tab to access the option of selecting a color:

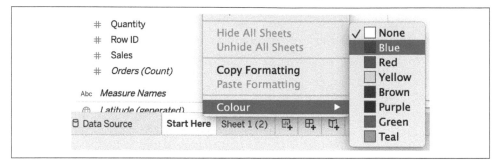

The color is then indicated underneath the worksheet name and is very handy when using filters—more on that later in this chapter:

Now that you have your first worksheet created, you're probably wondering what everything means in this interface.

2.2 The Tableau Workspace

The *Tableau Workspace* is the place where you build your charts. It uses a drag-and-drop principle, but you need to understand what and where to drag and drop. The interface has several areas that need explaining:

1. The first area that is brought to our attention is section A: the toolbar. The toolbar is the place to go to access commands, navigation, and additional analysis. It can also be used to control the formatting at different levels.

2. Section B is known as the sidebar. This contains the Data pane and Analytics pane (see Recipe 2.3).

3. Section C is a mixture of shelves and cards. This is where we will be adding data to create the views (see Recipe 2.7).

4. Section D is called the view, which is the canvas on the workspace. It is where the charts will be visualized. This will come to light in Chapter 3.

5. Section E is Show Me. The Show Me panel is used to show chart types suitable for the data you are using (see Recipe 2.8).

2.3 The Data Pane

Section B in the Tableau workspace is the sidebar, which contains the *Data pane*.

Problem

You want to understand what the Data pane is.

Solution

On the left side of the workspace, you'll see the Data tab. The Data pane is where all of your data source fields are stored. They are currently sorted by Tables (Customer Details, Monthly Targets, and Orders).

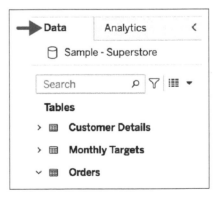

You can change the way these fields are grouped by clicking the drop-down menu next to the search bar:

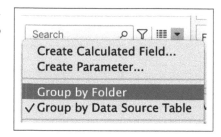

Discussion

The Data pane is the place where all your fields are stored from the data source. This will also be where any saved calculations are stored.

With the option of grouping by folder, you can create your own cut of the Data pane. For example, right-click any field and choose to Folders > Create Folder:

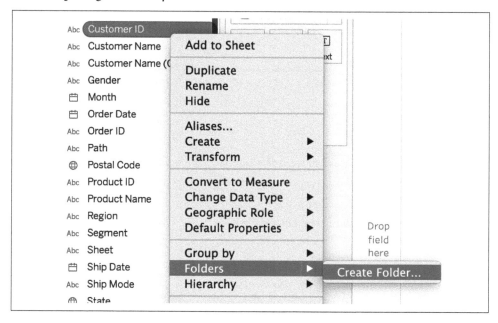

This allows you to group your data in a way that is meaningful to you.

However, you might be wondering what the faint gray line indicates in the list of fields:

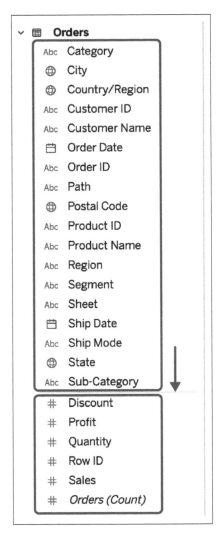

2.4 Dimension Versus Measure

Problem

You want to understand the difference between dimensions and measures.

Solution

1. Use the faint gray line as a guide. The fields above the gray line are dimensions:

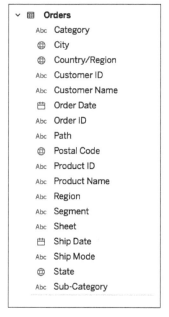

2. Now look at the fields below the gray line; these are measures:

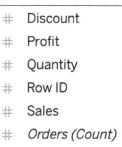

Discussion

Dimensions are fields that are qualitative, categorical data types, such as Colors, Names, and States. We will use dimensions to segment our views, and they define the level of details.

Measures, on the other hand, are quantitative data types or values you can measure, like Height, Weight, and Sales. Measures can have an infinite value and can be aggregated (summed or averaged).

You might be thinking, blue and green mean dimensions and measures, respectively; unfortunately, that is not entirely true, and we need to introduce discrete versus continuous.

2.5 Discrete Versus Continuous

Problem

You want to understand the difference between Discrete and Continuous.

Solution

1. Double-click a blue dimension. You will get headers within Tableau, which represent a finite range of dimensions:

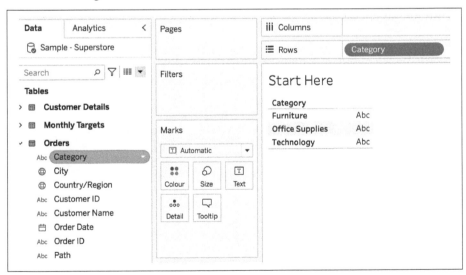

2. Double-click a green measure. You will get an axis within Tableau, which can have an infinite range:

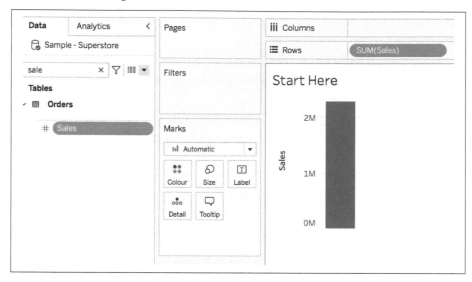

Discussion

The green field represents continuous data, and the blue field represents discrete. The definitions are as follows:

Continuous
  Forming an unbroken whole, without interruption

Discrete
  Individually separate and distinct

By default, Tableau uses discrete dimensions and continuous measures:

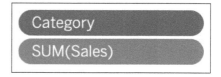

The other two options are discrete measures and continuous dimensions.

To understand the difference, we are going to use the SUM(Sales) field from the solution. If you right-click and select Discrete, this will convert the field to a discrete measure, which ultimately gives you a header:

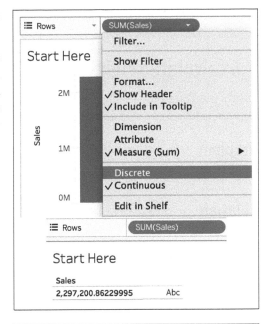

However, if you want to change the SUM(Sales) field to a continuous dimension, again you can right-click the SUM(Sales) field and select Dimension:

Notice now that you have many marks within the view. This is because it has now disaggregated the Sales value, so each mark represents an individual row in the data.

2.6 Calculated Fields

Inside the Data pane, you have the option to add calculated fields. Calculated fields contain calculations that could reference different fields in your data.

Problem

You want to create your first calculated field.

Solution

Click the drop-down arrow at the top of the Data pane and select Create Calculated Field:

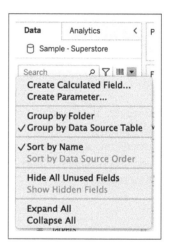

This will bring up a calculation window that allows you to start typing your calculations:

The search pane on the right-hand side allows you to see all of the available calculations, functions, and aggregations you can write in Tableau.

 You can open and close the calculation search pane by clicking the arrow on the right side of the dialog. More importantly, if you click any function within the calculation window, the definition and guidance will be shown in the search pane.

Discussion

Throughout this book, you will create a variety of calculated fields. You can create three types of calculations: Basic, Level of Detail, and Table Calculations. Tableau has two types of basic calculations: Aggregated and Non-Aggregated.

Working through an example will help you understand. We are going to calculate a Profit Ratio, which is Profit divided by Sales.

We'll start with Non-Aggregated, also known as a row-level calculation. This takes every row of data, performs the calculation, and then aggregates the value. When using the Profit Ratio example, the calculation would look like this:

This calculation will appear in the Measure section under the Orders Table. Notice that the icon next to the title has an equals sign. This shows it is a calculated field.

Reviewing this calculation on the following table shows the Profit Ratio for each Order ID and Product ID:

Category	Order ID	Product ID	Profit	Sales	Profit Ratio
Furniture	CA-2018-100090	FUR-TA-10003715	-88	502	-17.50%
	CA-2018-100678	FUR-CH-10002602	-18	317	-5.71%
	CA-2018-100706	FUR-FU-10002268	10	29	33.00%
	CA-2018-100916	FUR-TA-10004607	112	591	19.00%
	CA-2018-101462	FUR-FU-10000409	28	60	46.00%
	CA-2018-101560	FUR-FU-10003773	44	398	11.00%
	CA-2018-101602	FUR-CH-10004675	-22	763	-2.86%
	CA-2018-101931	FUR-BO-10001337	-36	617	-5.88%
	CA-2018-102295	FUR-CH-10001714	-18	121	-15.00%
	CA-2018-102652	FUR-FU-10000747	16	92	17.00%
		FUR-FU-10001918	13	33	39.00%

You will learn to build these visuals throughout the book.

However, we want to calculate the profit ratio per category—you'll notice that the profit ratio is above 100%:

Category	Profit	Sales	Profit Ratio
Furniture	18,451	742,000	8225.99%
Office Supplies	122,491	719,047	83177.06%
Technology	145,455	836,154	28838.70%

This is because the row-level calculation calculates the profit ratio for every individual row of data—in this case, for every Order ID and Product ID.

If we want to calculate the correct profit ratio, which can be used with any dimensions in the visualization, we need to convert it to an aggregate calculation. To do this, you can right-click the calculation in the Data pane and select Edit:

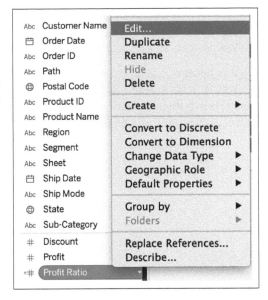

This will bring the calculation window back up. Here, we need to wrap both Profit and Sales in a SUM, to look like this:

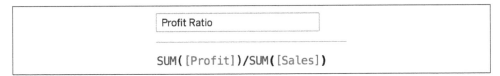

You cannot mix aggregate and non-aggregate; this error will appear if you aggregate only one measure:

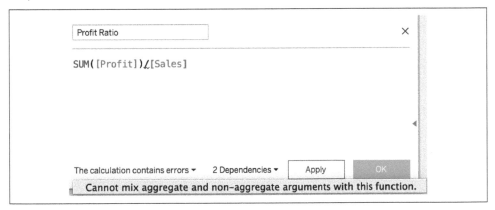

Once you save the calculation, since we have changed the calculation to an aggregate, it now moves position in the Data pane:

This is because calculated fields are dependent on which fields you use in a view. For example, you can use this calculation for Category, Segment, or any other dimension in your data.

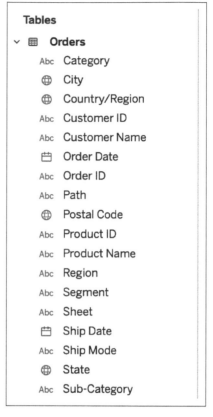

If you edit the same calculation from non-aggregate to an aggregate, and you have already used it in a view, you will get a red field because you have changed the aggregation of the calculation, which means you have to add it back to the views. Notice that it now says AGG(Profit Ratio) instead of SUM.

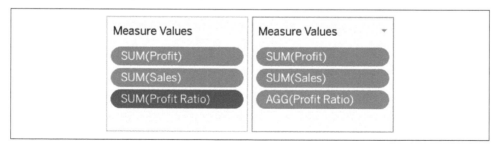

Now when we look at the calculation at the row level, it doesn't look any different:

Category	Order ID	Product ID	Profit	Sales	Profit Ratio (Non Agg)	Profit Ratio
Furniture	CA-2018-100090	FUR-TA-10003715	-88	502	-17.50%	-17.50%
	CA-2018-100678	FUR-CH-10002602	-18	317	-5.71%	-5.71%
	CA-2018-100706	FUR-FU-10002268	10	29	33.00%	33.00%
	CA-2018-100916	FUR-TA-10004607	112	591	19.00%	19.00%
	CA-2018-101462	FUR-FU-10000409	28	60	46.00%	46.00%
	CA-2018-101560	FUR-FU-10003773	44	398	11.00%	11.00%
	CA-2018-101602	FUR-CH-10004675	-22	763	-2.86%	-2.86%
	CA-2018-101931	FUR-BO-10001337	-36	617	-5.88%	-5.88%
	CA-2018-102295	FUR-CH-10001714	-18	121	-15.00%	-15.00%
	CA-2018-102652	FUR-FU-10000747	16	92	17.00%	17.00%
		FUR-FU-10001918	13	33	39.00%	39.00%
	CA-2018-102869	FUR-FU-10002456	5	17	28.75%	28.75%

But when we look at it at the Category level, you will see the correct profit ratio calculation:

Category	Profit	Sales	Profit Ratio (Non Agg)	Profit Ratio
Furniture	18,451	742,000	8225.99%	2.49%
Office Supplies	122,491	719,047	83177.06%	17.04%
Technology	145,455	836,154	28838.70%	17.40%

The type of calculation you need does depend on the use case and your data. Throughout the book, you will be creating a multitude of calculations.

2.7 Shelves and Cards

Within the Tableau workspace, shelves and cards play prominent roles in the building of visualizations. Tableau has several shelves and cards that can hold fields.

Rows and Columns

You use Rows and Columns in the workspace to create the visualization.

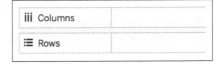

Problem

You want to understand the difference between Rows and Columns.

Solution

1. Add any discrete field to the Columns shelf. Tableau will then create a vertical separation within a table:

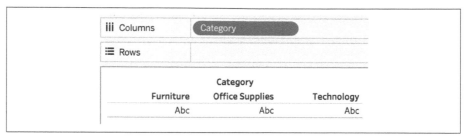

2. If you add another discrete field to the Rows shelf, Tableau creates a horizontal separation within a table:

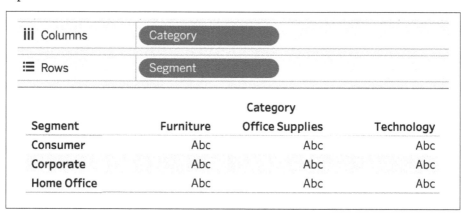

Discussion

The Rows and Columns shelves are fundamental to building any visualization. Creating the main types of visualizations are covered in Chapter 3 and beyond.

You can have multiple fields on rows and columns, which add more Rows and more Columns to the visualization.

Within the view on the workspace, you can also drag and drop fields to the relevant section, labeled, "Drop field here," to create rows and columns:

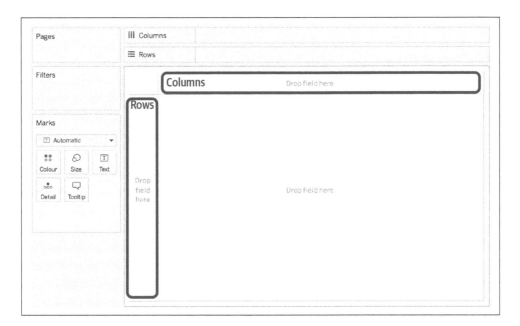

This solution uses discrete dimensions to create headers. If you use a continuous field on either Rows or Columns, you will create one row or column as an axis instead of a header:

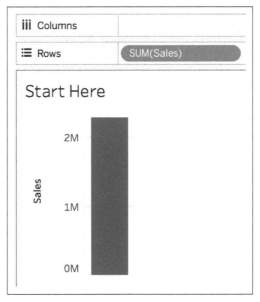

Marks Card

The *Marks card* is another fundamental part of building data visualizations in Tableau:

Problem

You want to understand what the Marks card does.

Solution

1. Drag and drop Segment onto Color:

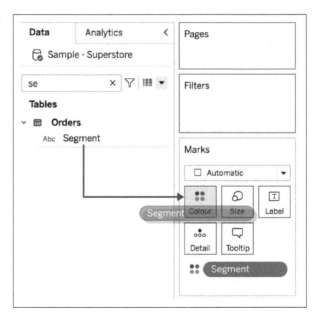

In the view, you will now see three colored squares, which represent the three segments:

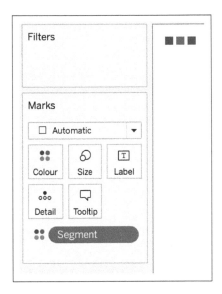

2. Adding a field to Color will also automatically show a color legend on the right:

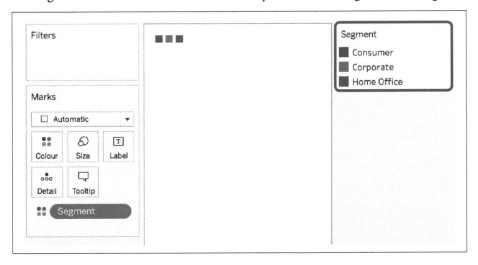

Hiding Show Me

If you cannot see your color legend, it might be hidden behind Show Me. If you click Show Me, Tableau will hide the Show Me menu, and you should see your color legend again:

3. The Marks card also controls the Mark Type. From the drop-down menu, click Automatic to see the types of Marks:

Discussion

When adding fields to the Marks card, you add level of detail and context to any visualization. The Marks card allows you to change the type of Mark you are visualizing. Each Mark Type has different properties available. Every type has the default five properties.

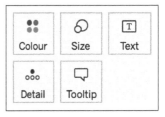

From Chapter 3 onward, I explain how to use each Mark Type and properties.

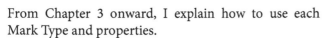

You can hide your color legend by right-clicking its drop-down menu and selecting Hide Card:

However, if you then decide you want the color legend back in view, you need to choose Analysis, then down to Legends > Color Legend (Category):

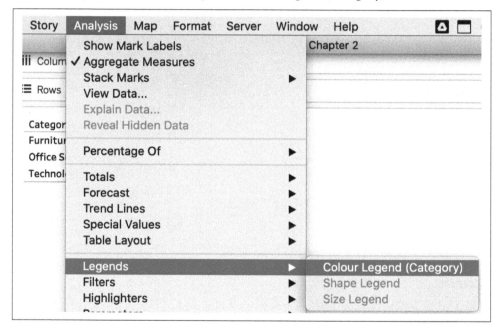

Filters Shelf

The *Filters shelf* appears above the Marks card, and is the place where you specify the data you want to include or exclude:

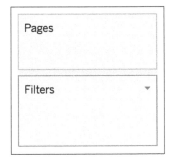

Problem

You want to allow the user to select the category for filtering.

Solution

1. Drag Category to Filters:

2. A pop-up box appears, which allows you to select any category you want to include or exclude. For this example, select the "Use all" option, and then click OK:

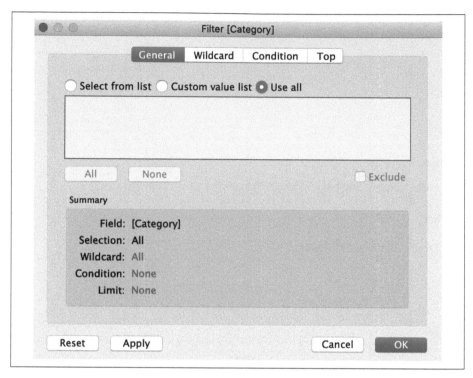

3. Now we have the Category field on the Filters shelf. To allow your users to interact with this filter, you have the option to show more details. Right-click Category on the Filters shelf and select Show Filter:

This shows the filter controls on the right-hand side of the canvas:

4. Currently the filter is set to "Multiple Values (list)," which means you can select multiple categories. To change it to a single-value list, click the drop-down menu on the filter control and select "Single Value (list)":

Your Filter control now looks like this:

Discussion

Filters are used to show certain cuts of the data, depending on what you have selected in the filter. You can apply more than one filter on a worksheet.

These filters can also apply to multiple worksheets in your workbook, especially if you have the filters on a dashboard. To apply this filter to another worksheet, right-click the Category field in the Filters shelf, and choose Apply to Worksheets > Selected Worksheets:

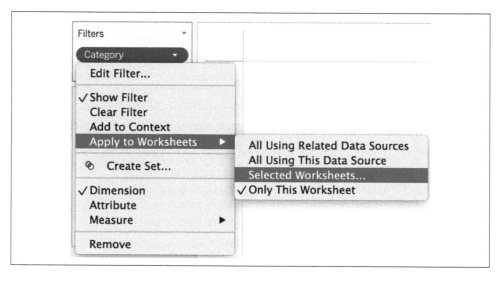

A pop-up box appears that allows you to select which worksheets you want the filter to apply to:

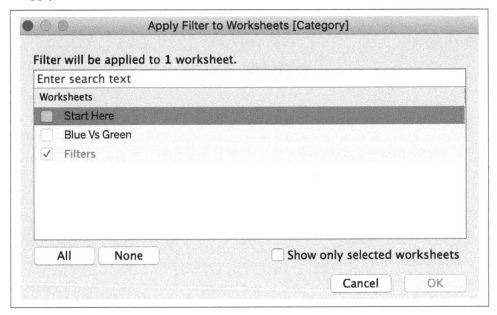

Here you can select the sheets within your workbook to apply this filter to. But notice that the Start Here sheet is highlighted in blue. In Recipe 2.1, I showed how to color worksheet tabs—the colors are great for grouping worksheets, but also when you want to apply filters to specific worksheets.

When adding fields to the Filters shelf, the fields can react in different ways depending on whether they are discrete or continuous (see Recipe 2.5).

In this solution, we used Category, which is a discrete dimension. This gives different options across the top.

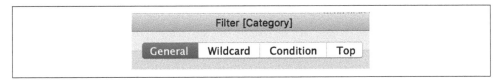

The Wildcard option is particularly useful if you have a lot of filter values. You can use this to search for a specific word within the field by using Contains, or you can search for words at the start or end of the fields:

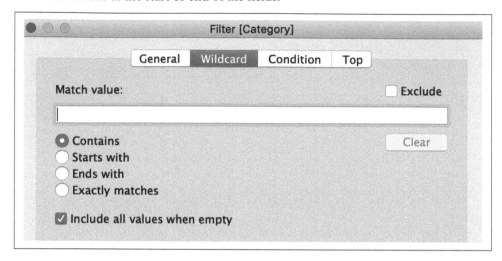

Condition and Top filters allow you to use a measure to determine which values to bring back; the condition and the selected category both have to return true.

For more information about Condition filters, see Recipe 14.2; for Top filters, see Recipe 3.3.

However, when you use a continuous field on the Filters shelf, like Sales, you'll have different options to choose from.

Continuous Filters

Problem

You want to see where the sum of sales is above a certain value.

Solution

1. Drag Sales to Filters:

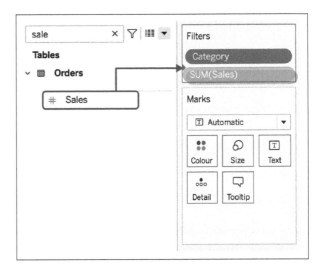

2. This will bring up a pop-up box for you to choose which aggregation you want. Select Sum and click Next:

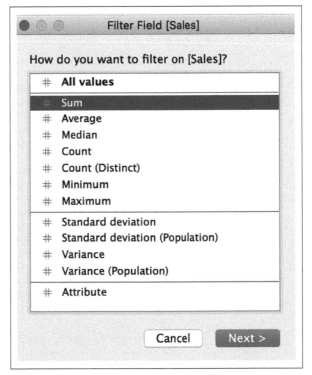

You'll be met with another pop-up box to determine the range of values:

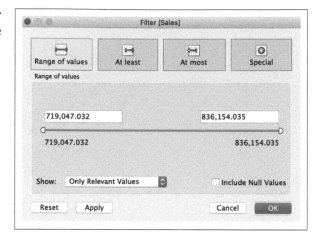

3. Once you click OK, the SUM(Sales) field will also appear on the Filters shelf:

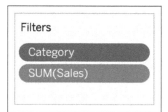

4. As with the Discrete example, we have to show the filter to let users interact with it. Right-click the SUM(Sales) field on the Filter shelf and select Show Filter:

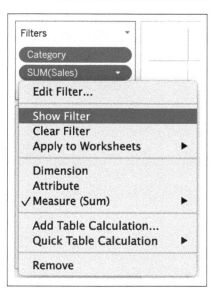

This shows the range of values within the filter:

Discussion

When the second pop-up box appeared, you had a few other options to select from:

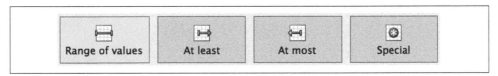

"At least" and "At most" allow you to specify either a minimum or maximum sales value. Special allows you to include or exclude null or non-null values that you may have in your data.

Now that we have added both the Category and Sales filters, we want these to be relative to each other. To do that, we can click the drop-down menu on any filter control and select Only Relevant Values:

Then when you change another filter, it will show only the relevant values associated with the previous selection. Changing the Sales value now returns only Technology:

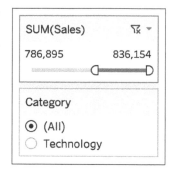

2.8 What Is Show Me?

The final section of the Tableau Workspace is *Show Me*:

Problem

You want to understand how to use Show Me.

Solution

Show Me allows the user to select an appropriate chart type depending on the data selected.

1. Select Category; then hold Ctrl (Command on Mac) while selecting Sales in the Data pane:

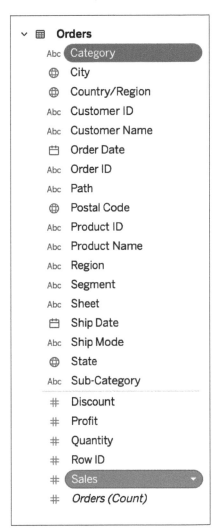

2. Now that you have selected the two fields, notice inside Show Me that the charts appear and that one has a box around it:

This is the recommended chart type for the data you have selected. Other charts not grayed out are also charts you can build with the data you selected.

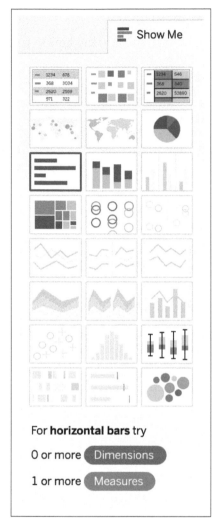

 If you have already built a chart, you can use Show Me to change to a different chart type, using the fields already in the view.

3. Select the recommended chart. Tableau will add the fields to Rows and Columns to create the chart:

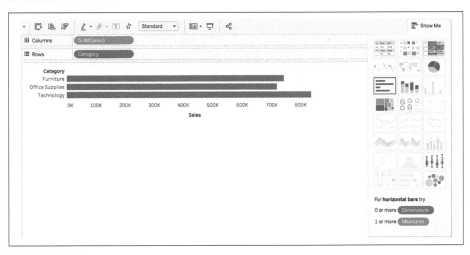

Discussion

Show Me is a great tool for helping you to start building different chart types in Tableau. However, to really understand Tableau and how to create the charts, I recommend using Show Me only if you really need to. The rest of this book will help walk you through the steps needed to create most of those charts within Show Me.

Because Show Me is open when you first go into the workspace in Tableau, you might think it has to stay there, but it does get in the way of the canvas. To minimize the Show Me box, click the Show Me tab:

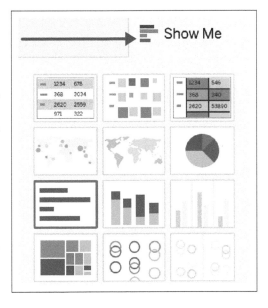

Now the canvas looks like this:

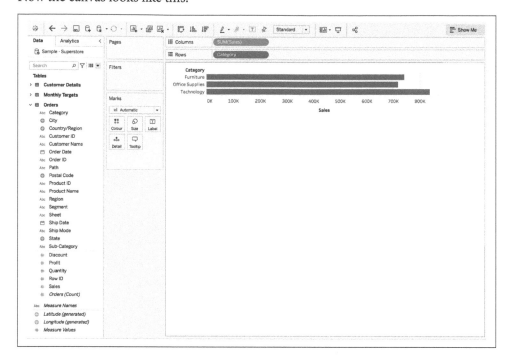

You have just built your first bar chart using Show Me, but as I mentioned, understanding how Show Me builds the charts is fundamental to mastering Tableau.

2.9 Formatting in Tableau

Before you start building visualizations, it might be helpful to start thinking about formats throughout the workbook.

Problem

You want to format the workbook to change all fonts.

Solution

1. In the menu, select Format > Workbook:

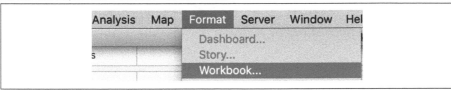

This opens a Format Workbook menu on the left:

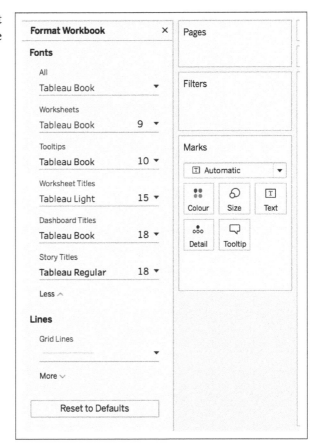

2. In this menu, you can format all text to a specific font. You can also remove all grid lines and axis rulers. Change an item, and Tableau will show a circle next to it to indicate something has changed:

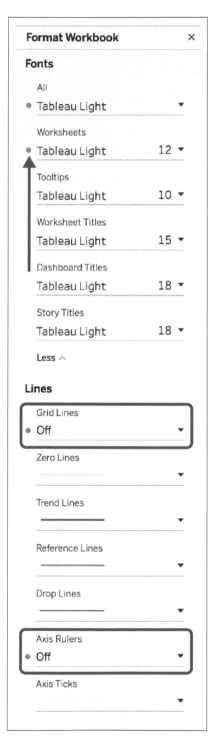

Discussion

Having the ability to format Tableau at the workbook level allows you to maintain consistency across worksheets.

If you don't want to format the entire workbook, you also have the option to format at an individual worksheet level. To format a worksheet, you can right-click in any whitespace on the worksheet and select Format:

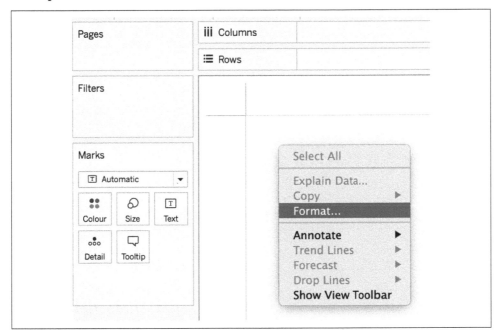

This will then bring up a Format pane on the right, which is slightly different from the workbook formatting:

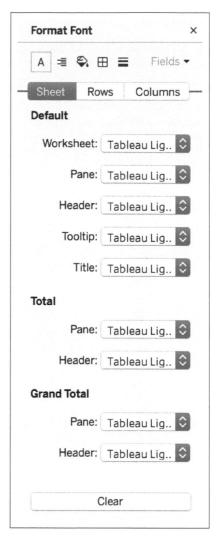

The worksheet formatting bar has five options across the top: Font, Alignment, Shade, Borders, and Lines:

You can also format at the Sheet level or, specifically, Rows and Columns.

If you have changed the format on one worksheet and want to add that to another worksheet, right-click a worksheet tab at the bottom and select Copy Formatting:

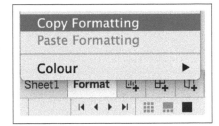

Once you have copied the formatting, the Paste Formatting option will be available for selection:

Summary

Chapter 2 has shown you the Tableau Workspace, which will help you get started with Tableau Desktop. You should now understand the differences between discrete and continuous, dimensions and measures, and aggregate and non-aggregate calculations.

Part I Conclusion

Part I has been about understanding the fundamentals of data and the Tableau Workspace before creating the visualizations. Part II will take you through some basic chart types to get started with Tableau Desktop.

Building a Foundation

CHAPTER 3

Bar Charts

A *bar chart* is one of the most commonly used chart types. It allows the human eye to compare differences between length or height, which is one of the preattentive attributes. A bar chart is split by categorical data and shows a continuous measure. There are two types of basic bar charts: horizontal and vertical. Bar charts have other variations, including stacked bar charts, diverging bar charts, and histograms.

3.1 Horizontal Bar Chart

Problem

You need to create a horizontal bar chart that shows, from highest to lowest, the continuous measure of Sum of Sales broken down by the categorical dimension of Sub-Category.

Solution

You have three options for this solution. Here is option 1:

1. Create a new sheet.

2. Using the Superstore sales data, double-click Sub-Category:

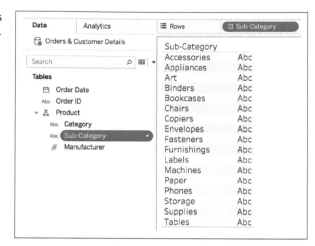

3. Drag the Sales measure to the Columns shelf:

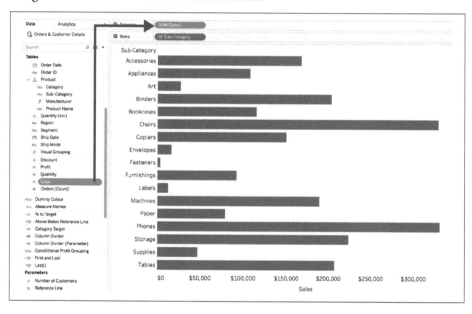

4. Hover over the axis to display a sorting bar icon. Click the sorting icon to create a descending bar chart:

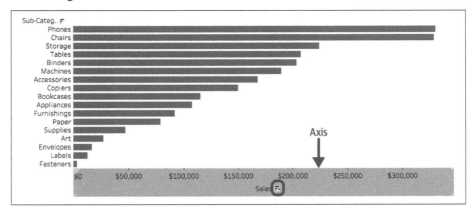

Here is option 2:

1. Create a new sheet.

2. Double-click Sales (1).

3. Double-click Sub-Category (2). This creates a vertical bar chart:

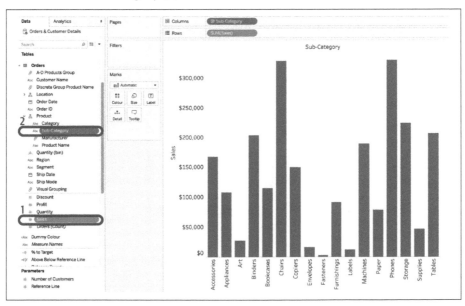

4. To convert this to a horizontal bar chart, click the "Swap Rows and Columns" button (1) in the toolbar.

Click the Sort Descending button in the toolbar (2), and you've converted a vertical bar chart to a horizontal bar chart:

And here is option 3:

1. Click Sales.

2. Hold Ctrl (Command on Mac) and select Sub-Category.

3. On the top-right side of the view, click the Show Me tab.

4. Click the chart that Tableau highlights as best suited for the dimensions selected in steps 1 and 3. The most suitable chart has a border.

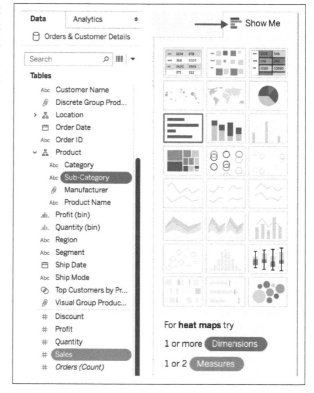

The chart type that Tableau has highlighted as the most suitable for the data might not always be the best choice or chart you are trying to create.

5. Another way to sort the data is to right-click the Sub-Category field in Rows and click Sort:

6. Change to sort by Field:

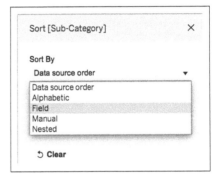

7. Change to Descending and confirm that the sorting is by Sales and the aggregation is Sum:

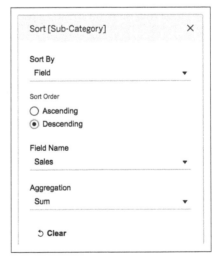

Discussion

In the solution provided, a horizontal bar suits the data more effectively than using the vertical bar chart. If you look back at option 2, where Tableau defaulted to building a vertical bar chart, you can see the Sub-Category headers are rotated vertically, meaning the reader would have to tilt their head to read the headers. With a horizontal bar, this allows the Sub-Category header to have more space and allows the reader to not shift position to read, therefore allowing faster interpretation of the chart. Vertical bar charts have their use cases too (see Recipe 3.10).

Adding a Constant Line

Bar charts are great for visualizing length as an indicator of a value, and they are also great for comparing whether a subcategory is above or below a *constant line*. A constant line is a static reference line that allows you to enter a single value for reference.

Problem

You need to have a reference line on your bar chart, to visualize where a constant value (e.g., a goal or target) compares to the sales of the subcategories.

Solution

1. Continue with one of your bar charts and click the Analytics pane, which is next to the Data pane on the left side:

2. Click, hold, and drag Constant Line to the view:

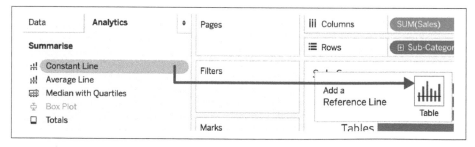

3. The Value box will automatically appear. Enter **100000**:

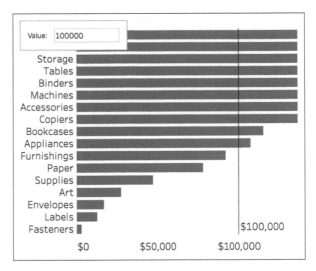

Discussion

A constant line is used for charts when you know you have a specific target or reference point. For example, each subcategory has a target of $100,000 in sales. The next section shows you how to input a number to control the constant line.

Changing to a User-Controlled Reference Line

In "Adding a Constant Line" on page 94, you learned how to add a constant reference line. A *user-controlled reference line* will allow the user to change the value of the reference line by using a parameter. Parameters are used to add interactivity to your worksheets and dashboards. They are covered in more detail in Chapter 15.

Problem

You need to change the constant number in the reference line on your bar chart, because the sales value will change when you add new data. This control allows you to change the reference line to suit.

Solution

1. Continue with one of your bar charts with the constant line (see "Adding a Constant Line" on page 94).

2. Right-click the axis and select Edit Reference Line:

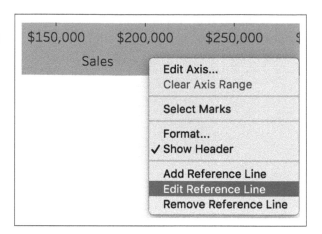

3. Click the drop-down option (1) and change the type to Sum (2):

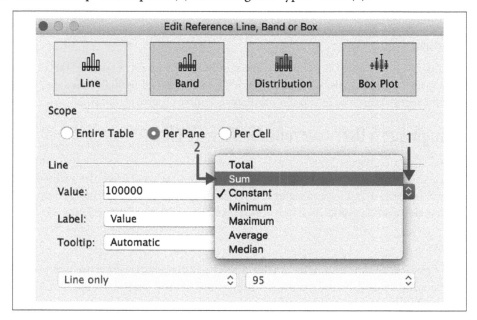

4. Then click the Value box and change it to "Create a New Parameter":

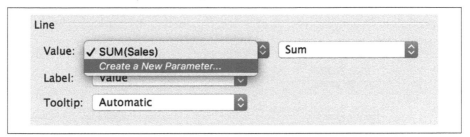

5. Rename the parameter and change the current value to **150,000**; then click OK:

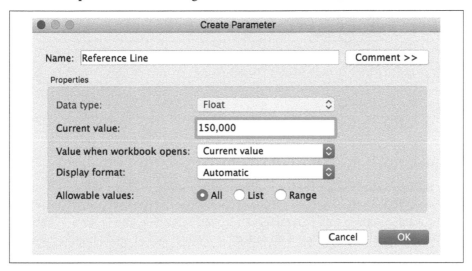

6. Customize the reference line by changing the Label drop-down to Value; then click OK.

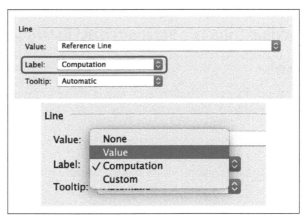

When you add a new parameter via a reference line, Tableau automatically shows the parameter control (1), which is the way you will be able to change the reference line. Whenever you change the parameter, the reference line will move accordingly.

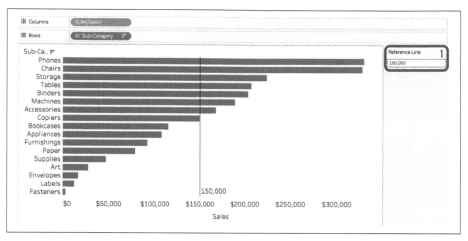

Discussion

Like the constant line, a parameter-driven reference line allows the user to input values to see which subcategories are above and below the reference line. The reason you might want to create this parameter-controlled reference line is for a What-If analysis. What-If analysis is seeing if an outcome changes based on changing the values. In this example, you could say, "What if Sales increased above $150,000; how many subcategories are above that value?"

Adding Color to a Parameterized Reference Line

Problem

You need to see whether each bar is above or below the controlled reference line on your bar chart. You can do this by using color.

Solution

1. Duplicate the previous sheet with the controlled reference line.

2. Create a calculated field and input this Boolean calculation:

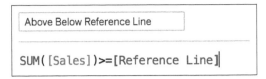

Above Below Reference Line

SUM([Sales])>=[Reference Line]

The purple text within a calculated field represents a parameter.

3. Add this calculation to the Color shelf. You can find the calculation under the list of fields in the Data pane:

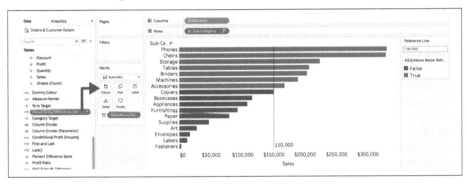

Discussion

Using color allows the user to quickly see which Sub-Categories are above or below the reference line. Tableau's automatic colors for a Boolean calculation are orange and blue, because they are the first two colors in Tableau's default color palette. Colors are very effective in data visualization because they're another example of a preattentive attribute, which allows us to process extremely efficiently.

But be careful when choosing colors. For example, some of your audience could be color-blind. In this case, it's important to ensure your colors are color-blind friendly. Otherwise, your audience might be able to see only shades of colors, depending on the type of color-blindness they have, which is determined by limits on someone's perception of the light spectrum.

The best example of this is using red and green. Those who are color-blind will see only shades of brown, and if you're using red to signify bad and green to signify good, the shades of brown could look very similar, so the good and bad implications will be lost. Tableau has included its own color-blind friendly color palette that I highly recommend using when you need to represent red or green.

Adding Average Lines

Constant lines and controlled constant lines are needed for user input, but what if you want to show the average per category?

Problem

You want to see a reference line for the average sales per category and per subcategory.

Solution

1. Double-click Category from the Data pane:

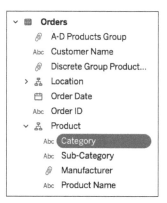

2. Click the plus sign button on the Category field to drill down to Sub-Category:

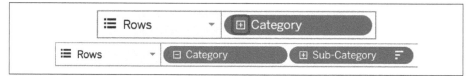

A plus symbol on a field means it is part of a hierarchy. A drilldown simply adds the next level of the hierarchy to the view. In the Data pane, you can see the hierarchy.

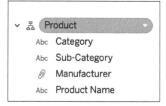

3. Drag Sales from the Data pane to Columns and sort descending by clicking the axis (see Recipe 3.1):

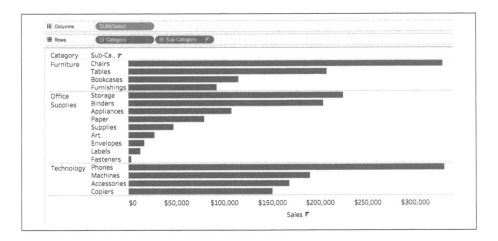

4. On the Analytics pane, hold and drag Average Line to the view. You have three options for the reference line (see the Discussion). If you want to get an average by category and you have a subcategory in the view, you will need to choose the Pane option. This will give an average of the subcategory sales within each category:

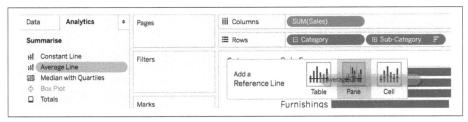

Your final view will look like this:

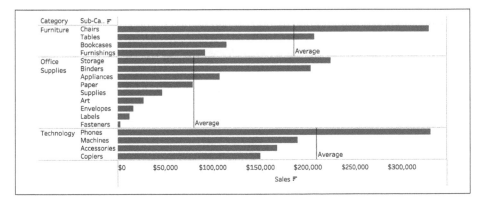

Discussion

Average lines are good if you have more than one field in your view. You have three options for average lines: Table, Pane, and Cell. Pane is mentioned in the solution just shown:

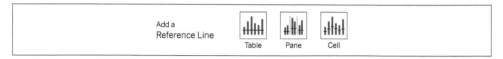

Table would give you the average of all subcategories not split by category:

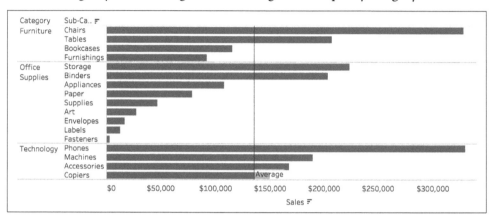

The Cell option would require a different level of detail; otherwise, it would return an average line per subcategory. To add additional detail and use the Cell average line:

1. Add [Order ID] to Detail.

2. Change the mark type to Circle.

3. Add a Cell average line.

This will give you the average sales per order within each subcategory:

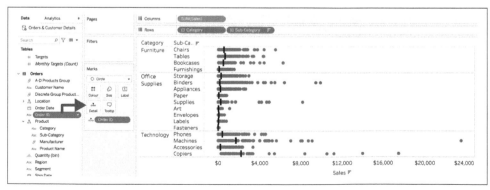

3.2 Grouping

Grouping in Tableau can be used as simple data preparation. For example, you might need to group values because of incorrect spellings or you may need to group people into teams. Groups are a manual process because they are based on data that you already have in your data source.

Problem

You need to manually group your data so that you can compare it to a newly created product group.

Solution

1. Right-click Product Name; then choose Create > Group:

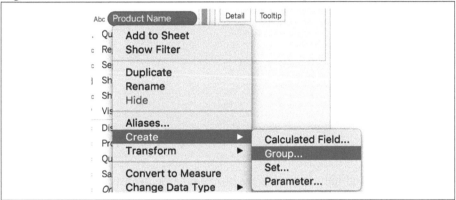

2. This allows you to create groups and select values to include in the group. Rename the group by editing the Field Name at the top to **A-D Products Group**.

3. Select the items from the list that you would like to group; then click Group. If you want to multiselect, press Shift as you select the first and last item:

4. While still highlighted, type in the box to rename the group to **A-D Products**:

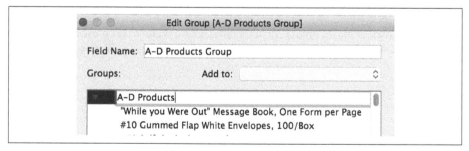

5. You can repeat this as many times as you need.

6. If you choose Include 'Other' this will group all the other ungrouped products into a group called Other. You might want to group things under the Other options, as it allows us to compare to the created groups against all the Other products in the group:

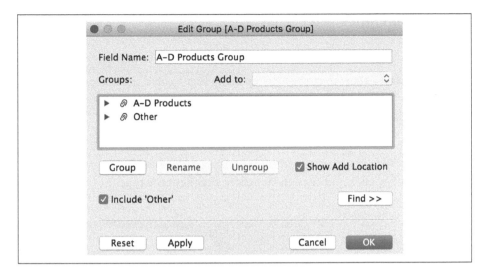

7. Once you have added all your groups, click OK.

8. Find the group in the Data pane and double-click it.

9. Drag Sales onto Columns:

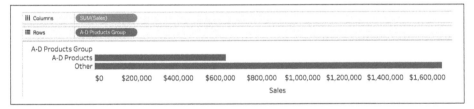

Discussion

Manual grouping is a fundamental part of Tableau and is not dynamic. For example, if you have new products beginning with A–D, then you would have to manually add them to the group. You can use one of three methods for grouping data:

- Visual grouping ("Visual Grouping" on page 106)
- Discrete grouping ("Discrete Grouping" on page 107)
- Conditional grouping ("Conditional Grouping" on page 111)

Visual Grouping

Problem

You want to use a chart to create a group using highlighted dimensions.

Solution

1. Create a horizontal bar chart with Product Name and Sum of Sales, sorted descending.

2. On the bars, click and lasso the marks to highlight the top 10:

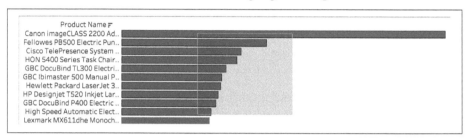

3. Hover over a highlighted bar and select the paper clip:

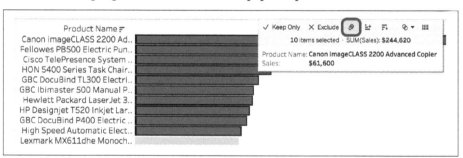

This creates a new group as a field in the Data pane, with the selected product names, and automatically adds it to Color on the Marks card:

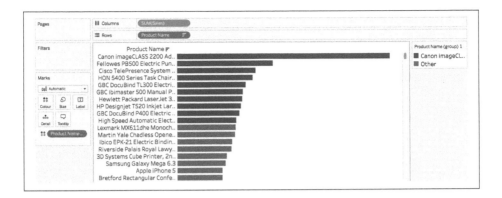

Discussion

Visual grouping is useful for seeing where marks are displayed within the view, particularly because Tableau automatically adds the visual group onto the Color property of the Marks card. Again, this group is not dynamic. As previously mentioned, visual grouping is a manual process. For example, if the data changes and one of the former products is no longer in the top 10 for highest Sales, you would have to repeat the steps mentioned to update the group.

Discrete Grouping

Problem

You want to group similar products together using the Product Name instead of their Sales position.

Solution

1. Create a horizontal bar chart with Product Name and Sum of Sales.

2. Hover over the start of the bar until you get a horizontal arrow. Drag right until you can see the full product name for 3D Systems Cube Printer:

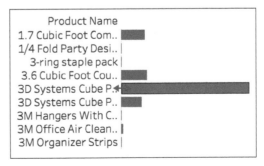

3. Press Ctrl, select both of the 3D Systems Cube Printers. Click the paper clip to group:

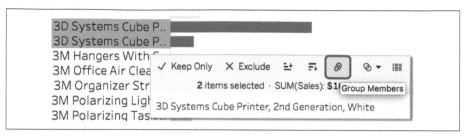

Notice that Tableau groups those two products into one bar and doesn't add the group to Color:

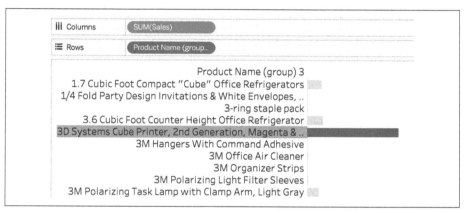

Discussion

This type of grouping is good for spelling mistakes or to combine dimensional fields. You'll notice that this type of grouping has added an "&" into the name of the product—we want the name to be *3D Systems Cube Printer*. To change the group name, right-click the dimension and select Edit Alias:

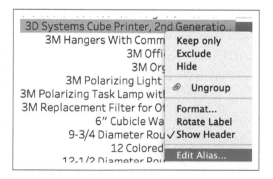

This allows you to change the name of the subgroup:

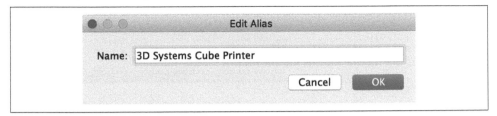

The grouped field will now look like this:

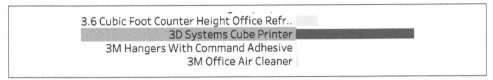

Finally, as Tableau has automatically created a field in the Data pane, it currently is called "Product Name (group) 2." Right-click and rename it **Discrete Product Name**:

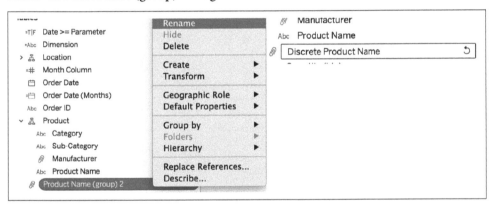

Now that you have this new group of products, you might not want your users to use the old Product Name field, but to use this new group going forward. First, you can replace the references of the Product Name field to this new group.

Right-click Product Name and select Replace References:

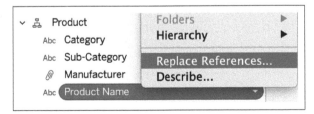

Find the field you want to replace it with; then click OK:

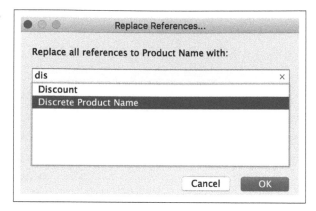

You'll notice that in the Data pane, Tableau has replaced Product Name with Discrete Product Name inside the hierarchy:

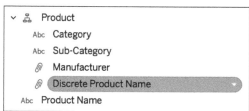

Second, you might want to hide the original Product Name field. To do this, right-click the Product Name field and select Hide:

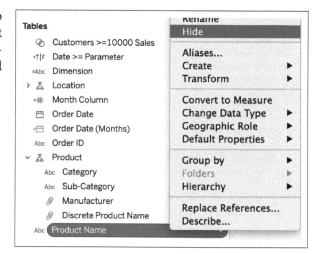

The field is now hidden from the Data pane:

To get back any hidden fields, click the drop-down next to the search box within the Data pane and select Show Hidden Fields:

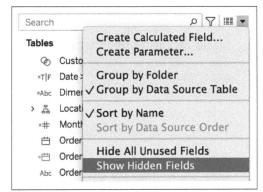

This will now show the hidden fields as light gray in the Data pane:

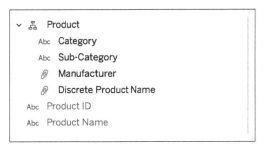

To unhide a field, right-click and select Unhide:

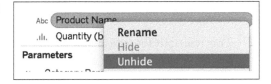

Conditional Grouping

Problem

You want to see three groups based on whether the profit of a subcategory is greater than or equal to $25,000, greater than or equal to 0, or below 0. This can be formed using a conditional IF statement.

Solution

1. Create a horizontal bar chart using Sub-Category and Sum of Profit, and sort descending by Sum of Profit (see Recipe 3.1).

2. Create a new calculated field and use the following IF statement (more on this in the Discussion). The values can be changed to suit. Click OK:

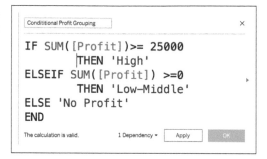

```
IF SUM([Profit])>= 25000
        THEN 'High'
ELSEIF SUM([Profit]) >=0
        THEN 'Low-Middle'
ELSE 'No Profit'
END
```

The calculation is valid.   1 Dependency ▾   Apply   OK

3. Drag the Conditional Profit Grouping calculation to Color on the Marks card.

This will split the subcategories into their respective conditional categories based on the IF statement you created in step 2:

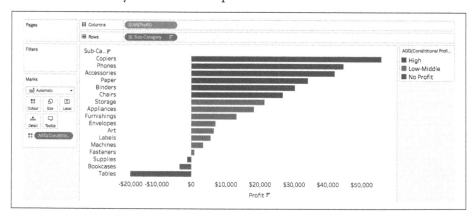

Now that you have this conditional calculation, you can also add it to the Rows, Columns, or Filters shelf to add additional detail to the chart or to restrict to a specific condition.

Discussion

This type of grouping is good if you have constant targets. You can create as many or as few conditional groups as you like. A simple IF statement works with the following logic:

```
IF this THEN that ELSE Something else END
```

The ELSE element of the calculation is optional and is a catchall for the values that don't follow the condition:

```
IF SUM([Profit]) > 0
    THEN "Greater Than 0"
ELSE "Less Than 0" END
```

If you need three or more conditions, you should add an ELSEIF before the ELSE. You can have many ELSEIF expressions in a conditional statement.

```
IF SUM([Profit]) > 0
    THEN "Greater Than 0"
ELSEIF SUM([Profit]) > −10000
    THEN "Greater Than −10000"
ELSE "Less Than 0" END
```

When using an IF statement, you need to always finish the statement with an END.

Should you need another example of a conditional statement, Tableau has an example built into the calculation window. Click the right arrow to expand the calculation window:

```
IF SUM([Profit]) > 0
    THEN "Greater Than 0"
ELSEIF SUM([Profit]) > −10000
    THEN "Greater Than −10000"
ELSE "Less Than 0" END
```

In the search box under the All drop-down menu, type **IF**. Click the third one down to get the example and description:

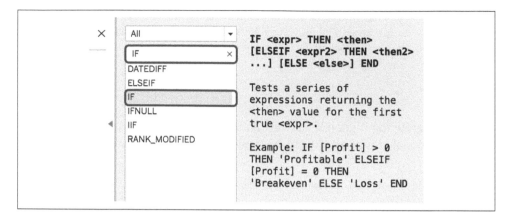

Using the search box within calculations is a great way to recap particular calculations within Tableau.

3.3 Top N Filter

Filters, as mentioned in Chapter 2, are useful for showing a subset of the data. A Top N filter (e.g., the top 10) allows you to show a subset of data based on a measure. This can be used to help with a lack of space in the visualization or because you want to focus on this top *N*.

Problem

You want to see only the top 20 customers by the Sum of Sales.

Solution

1. Create a horizontal bar chart using Customer Name and SUM(Sales), and sort descending by Sales.

2. Right-click the Customer Name field on Rows and select Filter:

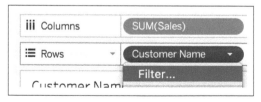

3. Select the option "Use all":

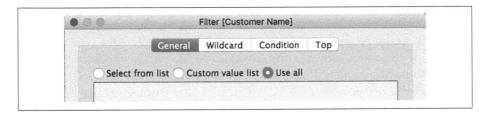

4. In the top bar, choose Top and select "By field." You want to select Top 20 by Sales and Sum. Click OK:

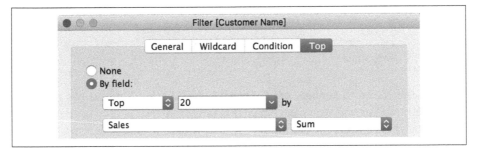

 The options inside the Top filters are all changeable.

This displays the top 20 customer names by the Sum of Sales:

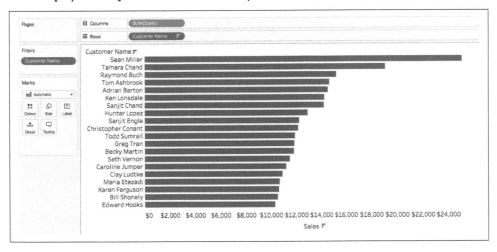

Discussion

Top N filters can be used with parameters, which would allow the user to change the number of customers shown in the selection. To get the filter options back up, click the Customer Name field that is on Filters and select "Edit filter." Click the drop-down box where you entered 20. Click "Create a New Parameter" (see "Changing to a User-Controlled Reference Line" on page 95):

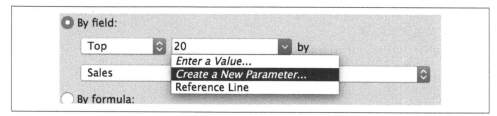

Change the Name to **Number of Customers**. Notice that Tableau has automatically chosen a range for you. You can change this range depending on your data and add in a Step size. Let's change the Minimum and Step size to 5. Click OK:

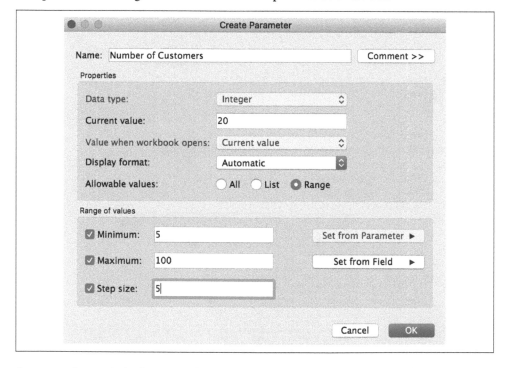

Once you have clicked OK to change the filter, the parameter control automatically displays on the right side of the worksheet, and you can click through the different numbers of customers:

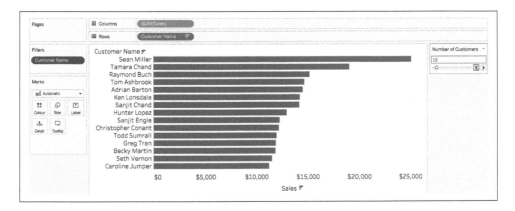

3.4 Stacked Bar Chart

Problem

You want to see total sales split by gender and subcategory.

Note that for the purpose of simplification, this dataset treats gender as a binary field.

Solution

1. Create a horizontal bar chart using Sub-Category and Sum of Sales, sorted descending by Sum of Sales.

2. Add the dimension (Gender) from the Customer Details data to Color on the Marks card:

3. To change the colors, click Color on the Marks card and Edit Colors:

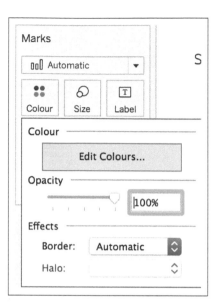

4. To adjust the colors, you need to select the data item on the left and then pick a color from color palettes on the right. If you double-click the data item, you will get the color pick option:

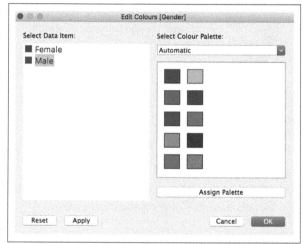

5. Once you have selected the colors you want, click the OK button, and this will be your final view:

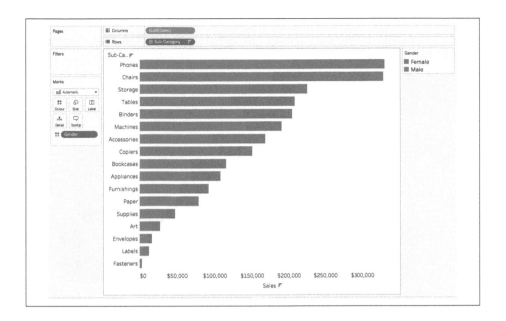

Discussion

Stacked bar charts are useful for showing potential relationships in your data and to display an extra level of detail. To enhance this chart type, you might want to add labels to each gender and a total per subcategory.

To get the labels per gender, click this Text icon button in the toolbar:

The view is broken down by gender; you might want to see the overall total value per subcategory. For this, you will need to add a reference line.

To add a reference line, right-click the Sales axis and click Add Reference Line. Change the scope to Per Cell, change the value aggregation to Sum and change the label to Value. Click OK.

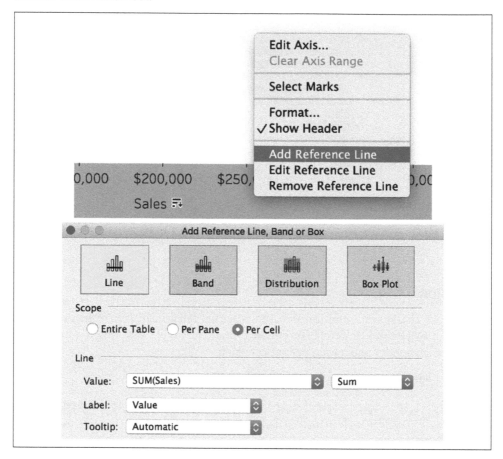

Notice that two labels are to the left of the reference line:

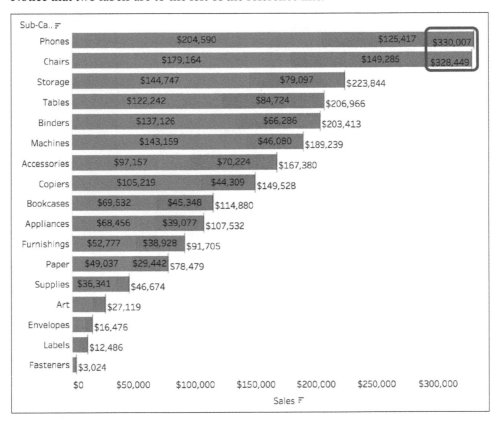

To right-align all the reference labels, right-click the reference line and select Format. Change the alignment to horizontal right-align and vertical middle-align.

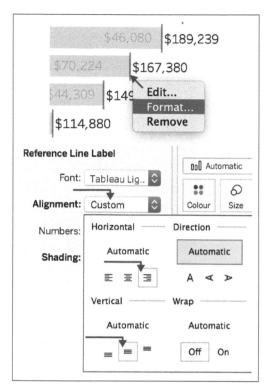

The chart now shows the values per gender and also the total value per subcategory:

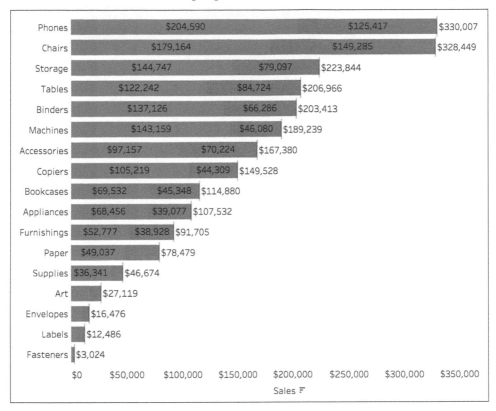

You can also change the reference line color to make it invisible. You can do this within the Format pane. Change the line color to None:

That makes the reference line disappear in the view:

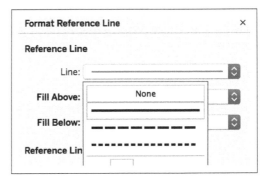

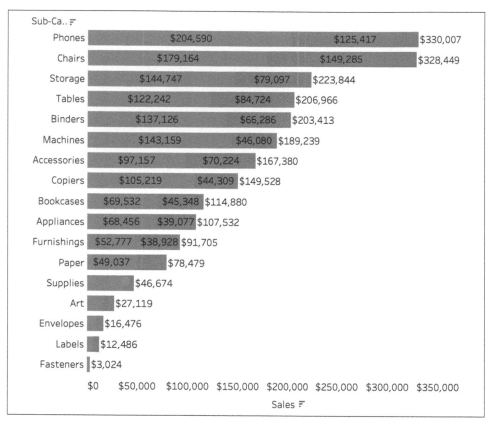

3.5 100% Stacked Bar Chart (Percent of Total)

Stacked bar charts are useful for showing contributions of each dimension of data, which is known as part-to-whole relationships. The drawback, however, is that 100% stacked bar charts don't allow for a comparison *across* the subcategories for each gender.

Problem

You want to see what percentage each gender buys from each subcategory by using "Percent of Total."

Solution

1. Create a horizontal stacked bar chart, with Gender on the Color property of the Marks card, by sum of Sales, sorted descending by Sum of Sales.

2. Right-click SUM(Sales) in Columns and choose Quick Table Calculation > "Percent of Total":

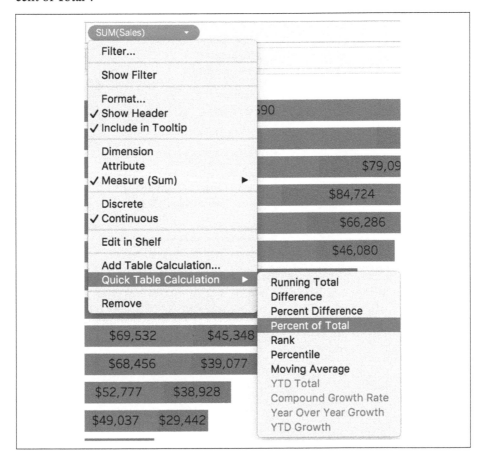

3. Tableau automatically performs the table calculation "Table (down)," which computes using the fields on the Rows shelf and on the Detail shelf. In this example, Table (down) means Tableau will compute the percent of total per gender per subcategory. For our use case, we want to compute using just gender.

4. Right-click SUM(Sales) in Columns and select Compute Using > Gender:

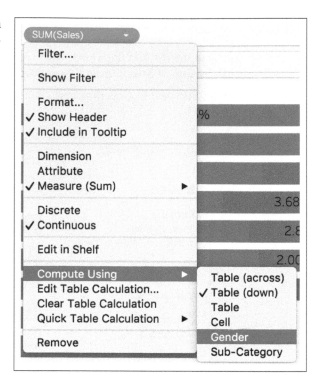

5. If you don't have percentage labels per gender, click Label on the Marks card and click the "Show mark labels" option.

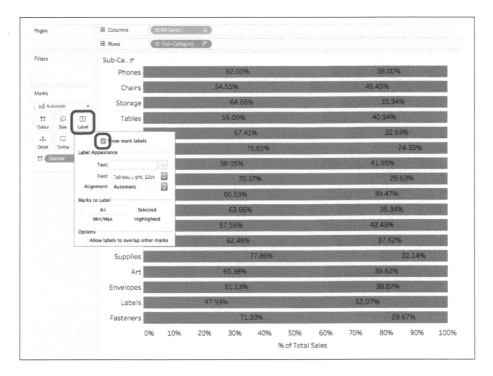

6. Tableau defaults percentages to two decimals, but this might be too precise. Right-click the SUM(Sales) field in Columns and choose Format:

7. On the Format pane, click Numbers, then Percentage, and finally, change to 0 decimal places:

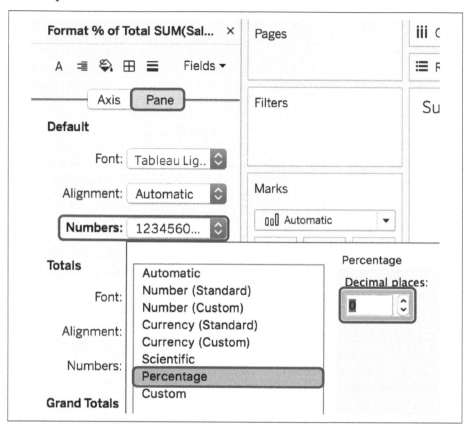

This is what your final view should look like:

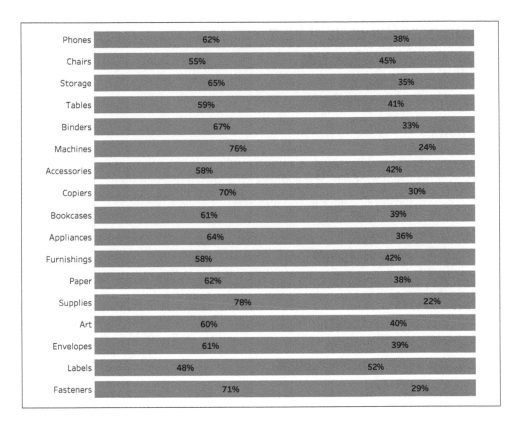

Discussion

100% stacked bar charts are ideal for comparing segments within each bar. In our example, the Percent of Total allows the user to see the percentage of each gender regardless of the total value for that subcategory. In reference to the solution, the chart allows you to see that more females buy from the Labels subcategory than males, and males buy more from Supplies than females. This type of chart can add deeper analysis.

When you see an upward triangle on a green field, that means the field has been changed to a Table Calculation:

Table Calculations are dynamic, and compute using the data you have in your view.

The stacked bar chart is just one way of showing part-to-whole relationships. Others include treemaps (see Recipe 8.1) or pie charts (see Recipe 11.13).

3.6 Discrete Bar in Bar Chart

Problem

You want to easily compare the sum of sales between male and female customers.

Solution

1. Create a horizontal bar chart showing Sub-Category and Sum of Sales, with Gender on the Color property of the Marks card, sorted descending by Sum of Sales.

2. Press Ctrl and click Gender to add it to Size.

 Pressing Ctrl duplicates the field you are dragging.

3. Tableau automatically stacks measures and values on top of each other because Tableau has Stack Marks on automatically. For this example, we want to compare males to females, which requires turning Stack Marks off. To do this, go to Analysis on the menu and choose Stack Marks > Off:

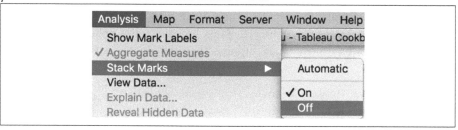

This creates a bar in the bar chart to show the difference between male and female sales.

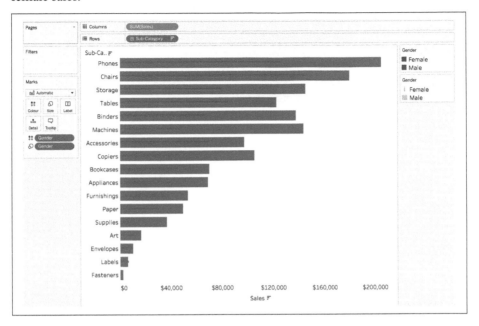

4. When you add a dimension or measure to Size, a size legend will appear on the right-hand side of the view. Change the size sort order of the gender by dragging Male above Female in the Size legend on the right side of the sheet:

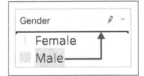

Then the chart looks like this:

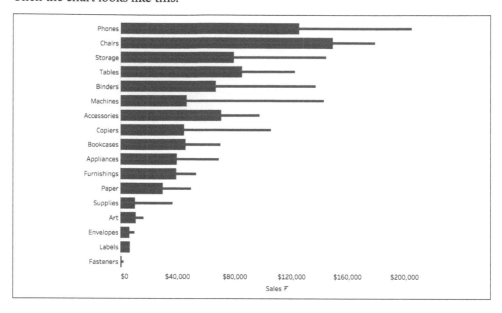

Discussion

This chart type is useful for comparing two discrete dimensions (male versus female) of two measures (budgets versus actuals; see Recipe 3.8).

3.7 Shared Axis

You might have a data set that has multiple measures in different columns. A *shared axis* will allow you to compare two or more measures on one axis. This is especially beneficial if the measures are on a similar scale.

Problem

You want to compare sales and targets side-by-side per category.

Solution

1. Create a horizontal bar chart using Category from the Monthly Targets tables and Sum of Sales:

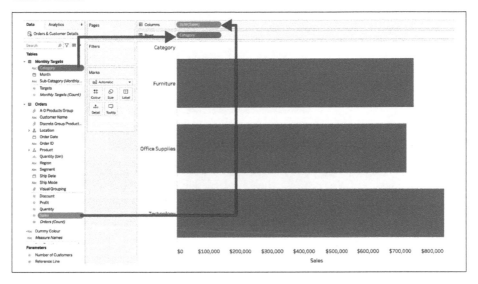

2. Click, hold, and drag the Category target on top of the Sales axis until you see two green rulers side by side. This means you are creating a shared axis:

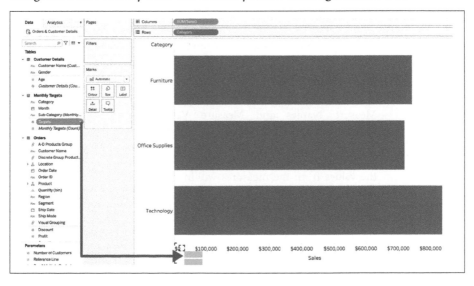

The final chart should look like this:

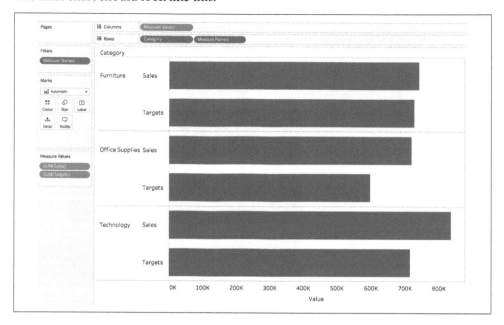

Discussion

Shared axes are useful if you have many measures with similar scales. Notice the new fields on Rows and Columns; Measure Names and Measure Values. Measure Names has also appeared on Filters, and a Measure Values shelf has appeared below the Marks card. These are all indicators that you have created a shared axis chart.

 Measure Names and Measure Values are Tableau-generated fields and come with every data set. They are used when trying to visualize multiple measures. Measure Values contains all the values from the measures within your data source in one single field. Measure Names contains the names of all the measures in your data.

3.8 Shared Axis Bar in Bar

Recipe 3.7 showed us how to create a shared axis, which is useful to show many measures on one axis. When trying to compare only two measures, a shared axis bar-in-bar chart might be more useful.

Problem

You want to show sales per category against a target.

Solution

1. Create a bar chart using Category from the Monthly Targets table, Target (Recipe 3.7), and Sum of Sales; or duplicate Recipe 3.7. This creates a side-by-side shared axis chart.

2. Move Measure Names from Rows to Color. Press Ctrl and duplicate Measure Names to Size:

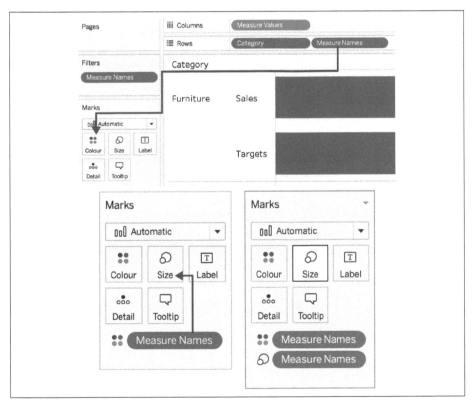

3. As mentioned in Recipe 3.6, Tableau automatically stacks measures and values on top of each other, because Tableau automatically stacks the marks. For this example, you will need to unstack the marks. From the menu, choose Analysis > Stack Marks > Off.

4. Shared axis bar-in-bar charts need to have a clear focus, and it is recommended that you mute the target color to gray and have the sales be a color of your choice for easier viewing. You will then be left with this chart:

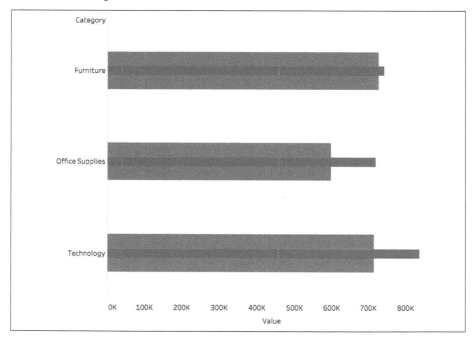

5. Adjust the sizes using the Size legend:

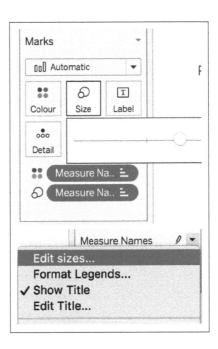

Discussion

As mentioned in Recipe 3.6, this chart type is useful for comparing either two measures (budgets versus actuals) or two discrete dimensions (male versus female). Be wary of using too many colors when creating charts like these as it can cause the user, interacting with the visualizations, to not know where to find the key insights. When comparing two measures, keep the color focus on the main value you want to highlight and use a gray for the background bar. If you're comparing discrete dimensions, make sure the colors don't clash and are distinguishable.

3.9 Bullet Chart

Problem

You want to compare category sales to targets and see references for 60% and 80% of the target.

Solution

Multiselect Category and Targets from the Monthly Targets table and Sales from the Orders table.

Click Show Me and choose the bullet chart (the bottom-middle chart).

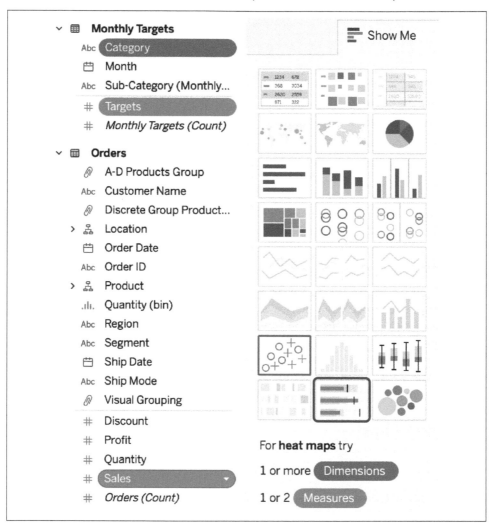

This will automatically create a bullet chart showing Sales versus Targets and 60% and 80% of the target:

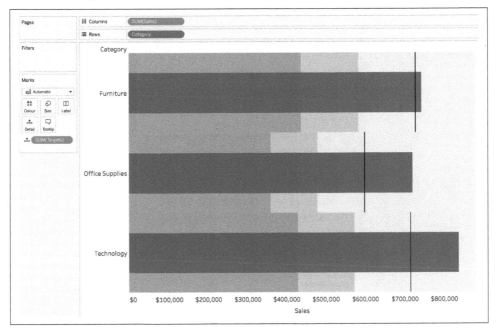

Discussion

The bullet chart is useful for comparing one measure against another, and it also provides a reference point for 60% and 80% of the target. This is useful when you refresh the data, as it shows progress toward the target.

If you find that Tableau has used the wrong reference line field, you can swap it by right-clicking the axis and selecting Swap Reference Line Fields. This will swap target and sales, meaning the target is now the bar, and the sales is the reference line:

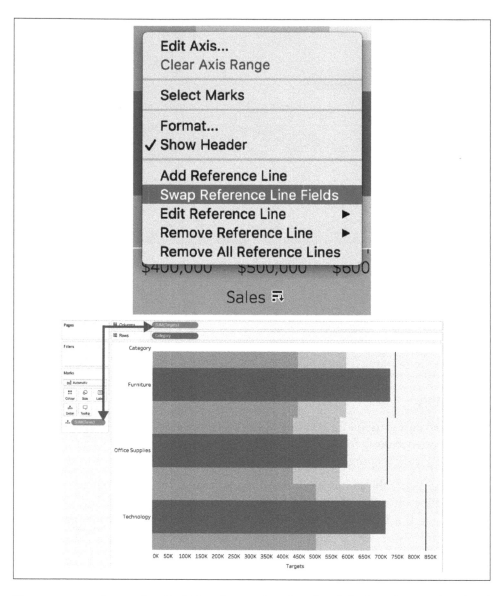

If you want to change the 60–80% reference band, right-click the reference band and then click Edit.

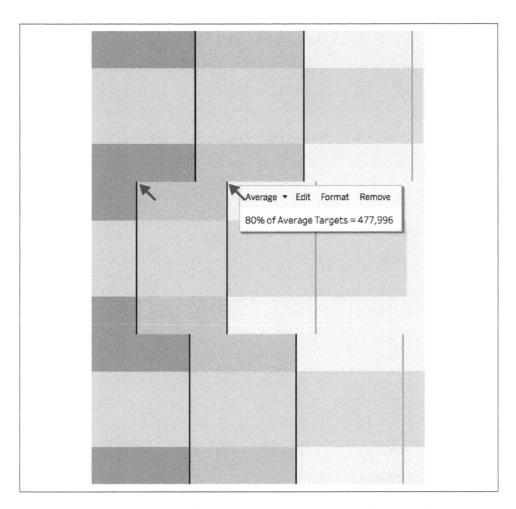

Here you can change it to different percentages; use percentiles, quantiles, or standard deviations; and you can also change the formatting of the reference band.

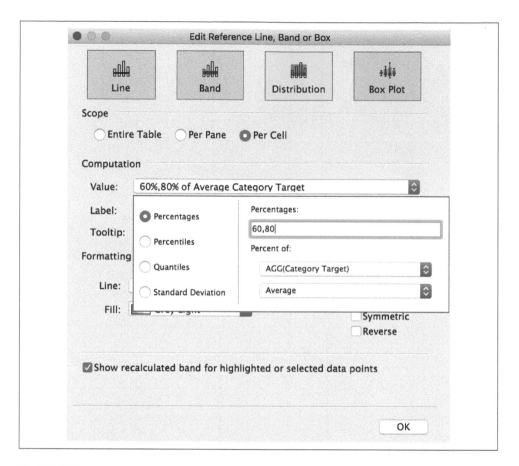

3.10 Histogram

A *histogram chart*, unlike a bar chart, shows the distribution and frequency of data.

Problem

You want to find the distribution of the number of orders for each quantity.

Solution

1. Select Quantity from the Orders table on the left. Click Show Me and then click the histogram chart.

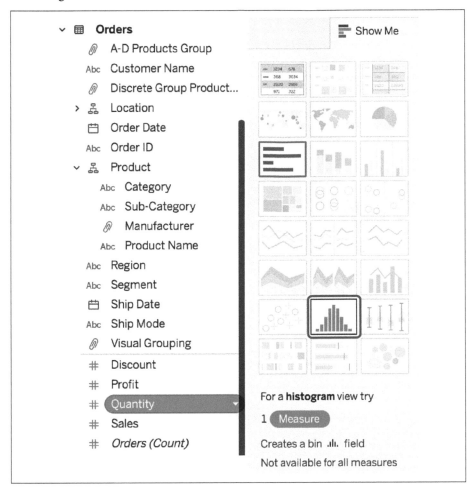

Tableau automatically creates a histogram to show the number of orders that have purchased N quantity and also automatically creates a new field called "Quantity (bin)":

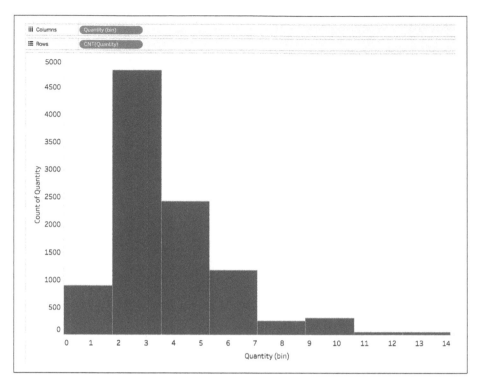

2. You can also change the size of the bins, by right-clicking "Quantity (bin)" and changing the bin size to 2.

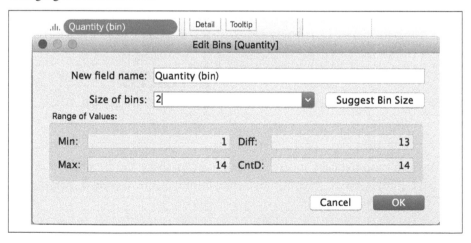

Discussion

This chart shows how many items are in the order. More than four thousand orders had greater than two but fewer than four items:

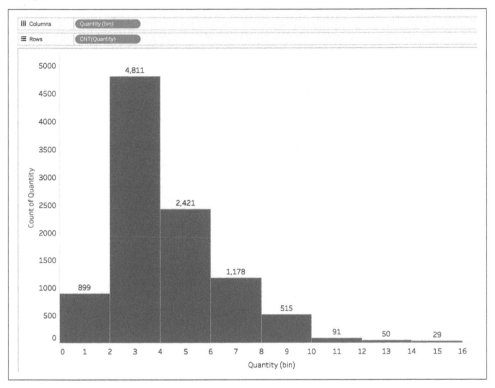

This type of chart shows the distribution in your data using equal bin sizes. The bin sizes can be static, as the solution provided, or you can use a parameter to let the user choose the bin size.

3.11 Soundwave

Bar charts are heavily used in data analytics. The *soundwave chart* is a version of the bar chart. It uses the standard bar chart approach as mentioned previously and requires an extra step to allow the same data to be replicated above and below. The soundwave chart can be used to enhance the visualization of volume.

Problem

You want to visualize the quantity of each order in the latest year of data.

Solution

1. Drag Order Date to Filters and select Month/Year. Select the "Filter to latest date value when workbook is opened" checkbox and then click OK.

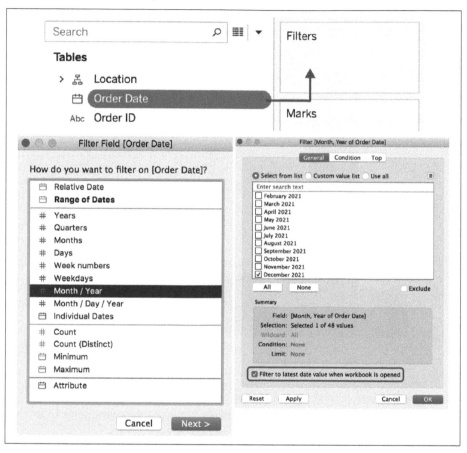

2. Double-click Quantity and double-click Order ID to create a bar chart.

3. Duplicate the SUM(Quantity) field on the Rows shelf. Double-click the second field and add "–" to the front. Then press Enter:

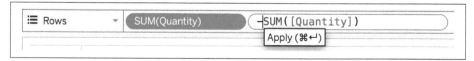

4. To format this, right-click the axis and uncheck the Show Header option. Then right-click the Order ID field on Columns and also uncheck the Show Header option:

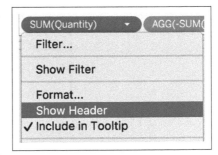

You now have a soundwave chart that looks at how much quantity was purchased by each order:

Discussion

Soundwave charts are great for showing volume. This type of chart doubles the extreme values. For example, you can see that this particular order (CA-2020-117457) has the highest value for quantity:

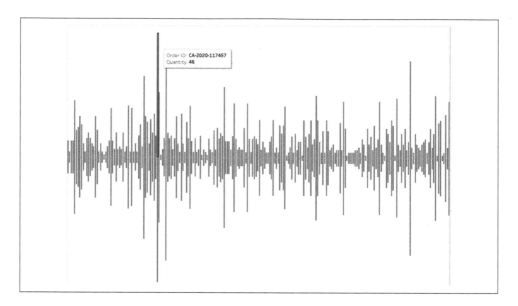

To make this chart more accurate, you should divide the quantity value by 2 for the positive and negative axis, so you have half the quantity on each side. To do this, double-click the field on the Rows shelf and add **/2** to the end, and repeat for the minus field:

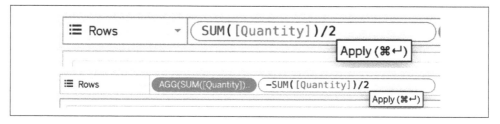

But now when you hover over, Tableau is giving the wrong value for the bars:

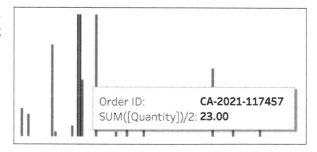

We can tell Tableau not to include these ad hoc fields in the tooltip by right-clicking the fields on the Rows shelf and deselecting "Include in Tooltip":

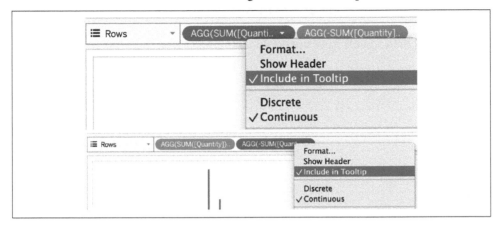

Then we can add the correct Quantity value to the tooltip by adding Quantity to Tooltip:

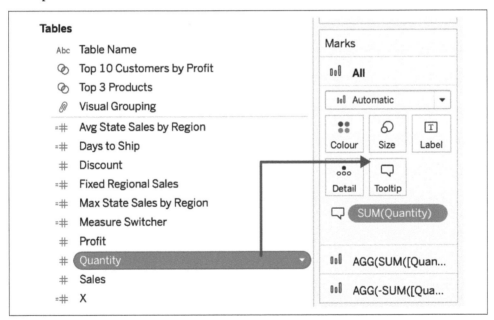

This chart is a different way of showing the same data and would suit call center data, where the length is the number of minutes on a call, or word counts. It is called a soundwave chart because it looks like an audio wave chart.

Summary

Throughout this chapter, you have learned about the most commonly used chart in data visualizations, the bar chart. Alongside this, I have also sprinkled many techniques that will help you on your journey of learning Tableau.

CHAPTER 4

Text

The Marks card features many options; one of those is Text. The Text property on the Marks card allows you to display text, which could be labels on a bar, text used to create tables, or text used as a mark type itself. This chapter will take you through some of those features, including tips and tricks for formatting.

4.1 Tables

Problem

You want a text table showing sales by category and year.

Solution

1. Double-click Category.
2. Double-click Order Date.
3. Double-click Sales:

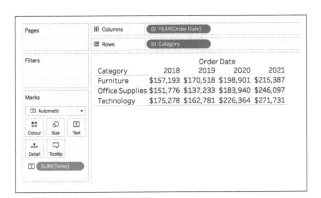

Discussion

A text table can be used in a variety of ways. The main use case is to show the specific values by the dimensions you have used. Tables are also great for sense checking the numbers and calculations when conducting analysis.

If you are using multiple measures, using Show Me to build a quick table might be easier. If you multiselect the fields from the Data pane, Tableau will make a table available in the Show Me menu:

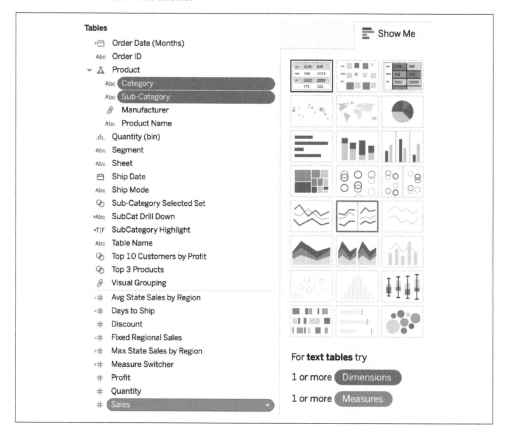

Finally, if you have already created a chart, you can right-click the Sheet tab and duplicate the sheet as a crosstab, which will give you the data in a table format:

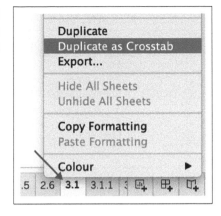

4.2 Adding Totals

Problem

You want to see the totals per category, per year, and grand totals.

Solution

1. Choose Analysis from the menu and then select Totals:

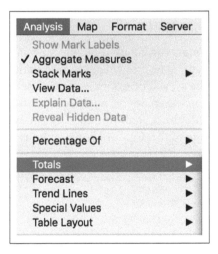

2. Choose Show Row Grand Totals:

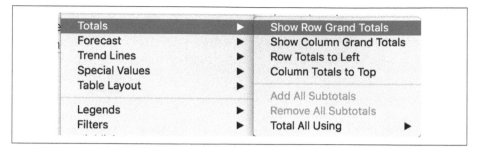

3. To add column grand totals, repeat steps 1 and 2, except choose Show Column
   Grand Totals.

Your final table should look like this:

Category	2018	2019	Order Date 2020	2021	Grand Total
Furniture	$157,193	$170,518	$198,901	$215,387	$742,000
Office Supplies	$151,776	$137,233	$183,940	$246,097	$719,047
Technology	$175,278	$162,781	$226,364	$271,731	$836,154
Grand Total	$484,247	$470,533	$609,206	$733,215	$2,297,201

Discussion

Totals can be used on almost every chart you create in Tableau. Tableau defaults the
aggregation of totals to Automatic, which is based on the aggregation of the field in
the view. For example, if you use an average aggregation, your totals will be an average, whereas you might want it to be the sum of the averages. However, you can over-
ride the default by choosing Analysis from the menu, then Totals, and then choosing
one of the Total All Using options:

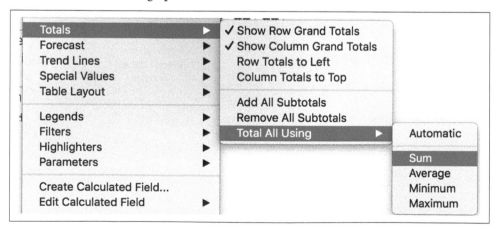

You can also add subtotals, depending on the level of detail in your view, which is set by another discrete field in the view. For example, this new table shows Category and Sub-Category sales by year, but you might want to see the subtotals per category:

Category	Sub-Categ..	2018	2019	Order Date 2020	2021	Grand Total
Furniture	Bookcases	$20,037	$38,544	$26,275	$30,024	$114,880
	Chairs	$77,242	$71,735	$83,919	$95,554	$328,449
	Furnishings	$13,826	$21,090	$27,874	$28,915	$91,705
	Tables	$46,088	$39,150	$60,833	$60,894	$206,966
Office	Appliances	$15,314	$23,241	$26,050	$42,927	$107,532
Supplies	Art	$6,058	$6,237	$5,961	$8,863	$27,119
	Binders	$43,488	$37,453	$49,683	$72,788	$203,413
	Envelopes	$3,856	$4,512	$4,730	$3,379	$16,476
	Fasteners	$661	$545	$960	$858	$3,024
	Labels	$2,841	$2,956	$2,827	$3,861	$12,486
	Paper	$14,835	$15,288	$20,662	$27,695	$78,479
	Storage	$50,329	$45,048	$58,789	$69,678	$223,844
	Supplies	$14,394	$1,952	$14,278	$16,049	$46,674
Technology	Accessories	$25,014	$40,524	$41,896	$59,946	$167,380
	Copiers	$10,850	$26,179	$49,599	$62,899	$149,528
	Machines	$62,023	$27,764	$55,907	$43,545	$189,239
	Phones	$77,391	$68,314	$78,962	$105,341	$330,007
Grand Total		$484,247	$470,533	$609,206	$733,215	$2,297,201

If you right-click on the Category field on the Rows shelf, you will get the option to add Subtotals:

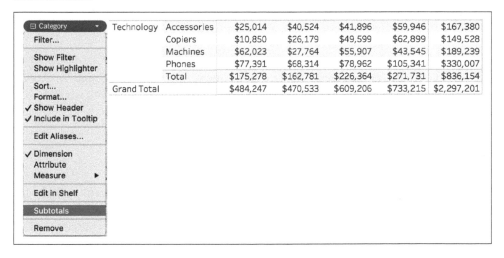

By default, Tableau puts the Totals and Subtotals to the bottom or right of any chart. We can move these Totals to the top or left by choosing Analysis from the menu, then Totals, and selecting either "Row Totals to Left" or "Column Totals to Top":

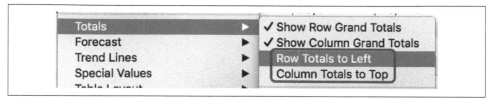

Top Tip 1: Adding Totals from the Analytics Pane

You can also add totals by using the Analytics pane. If you select the Analytics pane on the left, you will see Totals as an option; drag and drop that onto the middle of the view, and this will add both the Column and Row Grand Totals to the chart:

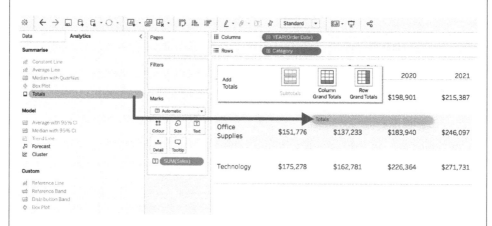

Using the Analytics pane totals allows you to choose either Column Grand Totals or Row Grand Totals. If you had an additional discrete field in the view, you would have the option to add subtotals as well.

4.3 Highlight Tables

Tables are good for seeing the actual numbers, and sometimes you want to make the table more intuitive and effective by adding color to transform your table into a *highlight table*. As previously mentioned, color is one of the preattentive attributes that allows the user to quickly interpret the chart.

Problem

You want to see the highest and lowest overall profit by subcategory and region by using color and text.

Solution

1. Double-click Sub-Category and double-click Region.
2. Drag Profit to Color on the Marks card:

Your current view should look like this:

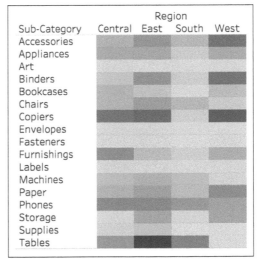

3. Finally, using Label on the Marks card, check the "Show mark labels" option to turn on the text labels to show the profit values:

Your highlight text table should look like the following:

| | Region | | | |
Sub-Category	Central	East	South	West
Accessories	$7,252	$11,196	$7,005	$16,485
Appliances	-$2,639	$8,391	$4,124	$8,261
Art	$1,195	$1,900	$1,059	$2,374
Binders	-$1,044	$11,268	$3,901	$16,097
Bookcases	-$1,998	-$1,168	$1,339	-$1,647
Chairs	$6,593	$9,358	$6,612	$4,028
Copiers	$15,609	$17,023	$3,659	$19,327
Envelopes	$1,778	$1,812	$1,465	$1,909
Fasteners	$237	$264	$174	$275
Furnishings	-$3,906	$5,881	$3,443	$7,641
Labels	$1,073	$1,129	$1,041	$2,303
Machines	-$1,486	$6,929	-$1,439	-$619
Paper	$6,972	$9,015	$5,947	$12,119
Phones	$12,323	$12,315	$10,767	$9,111
Storage	$1,970	$8,389	$2,274	$8,645
Supplies	-$662	-$1,155	$2	$626
Tables	-$3,560	-$11,025	-$4,623	$1,483

Discussion

The highlight table shows hotspots or cold spots in your data. For example, the Tables/East cell, which is dark orange, is the least profitable, and the Copiers/West cell, which is dark blue, is the most profitable. The ability to quickly pick out those values was enhanced by the use of color.

The automatic color palette Tableau has used is Orange-Blue Diverging. Diverging color palettes should be used where you have a crossover point, i.e., negatives to positives or above or below a target. If you have only positive or only negative values, you should use a sequential color palette. You can change the colors by clicking Color on the Marks card. The list of available colors depends on the type of values on Color.

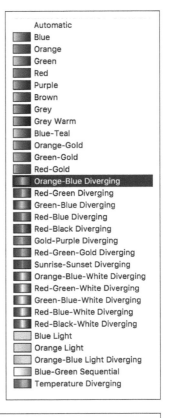

Top Tip 2: Custom Color Palettes

You can add your own custom color palettes. You do this inside your Tableau Repository, which is typically located in your documents. The repository contains a file named *preference.tps*. Right-click it and open it with a text editor. If you haven't edited this before, you will see just the following text:

```
<?xml version='1.0'?>
<workbook>
</workbook>
```

You can add three types of color palettes: regular, ordered-sequential, and ordered-diverging. Each color palette is used for different types of data.

Regular is used for categorical data and can be used with discrete dimensions. *Ordered-sequential* is mainly used for a single color ranging from light to dark and is available using continuous measures on the Color property of the Marks card. The final type is *ordered-diverging*, which has two ranges using the color intensity and a midpoint. This type is primarily used for continuous measures and for showcasing a crossover point within the values.

To start adding a color palette, you will need to add a `<preferences>` tag to your code, between the `<workbook>` tags:

```
<?xml version='1.0'?>
<workbook>
    <preferences>
    </preferences>
</workbook>
```

For every color palette you want to add, you need to open and close a `<color-palette>` tag. You also need to give it a recognizable and nonambiguous name, as this is the name that will appear when you are choosing colors. The name needs to be followed by one of the three types mentioned. Then another tag is needed for each color you would like in the palette, followed by the hex code of a color, as in this example:

```
<?xml version='1.0'?>
<workbook>
    <preferences>
        <color-palette name = "Name of Palette" type ="regular">
            <color>#000000</color>
            <color>#FFFFFF</color>
            <color>#111111</color>
        </color-palette>
    </preferences>
</workbook>
```

Once you have added your custom color palettes, you need to close and reopen Tableau for them to appear in the drop-down list of colors.

4.4 Rank Tables

As mentioned, tables are good for seeing precise numbers. When looking at values, you might want to see where those numbers sit in terms of a rank.

Problem

You want to see how each subcategory ranks for each region.

Solution

1. Build a table with Sub-Category on Rows, Region on Columns, and Profit on Text.

2. To change Profit into a rank, you will need to add a Quick Table Calculation. Right-click the SUM(Profit) field on the Text property. Select Quick Table Calculation > Rank:

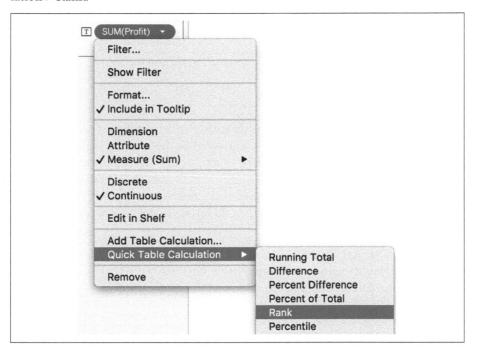

By default, Tableau computes the rank by region (table across); that is, for every subcategory, it ranks regions based on profit:

		Region		
Sub-Category	Central	East	South	West
Accessories	3	2	4	1
Appliances	4	1	3	2

3. To change the direction Tableau calculates the rank, right-click SUM(Profit) on the Text property, which now has a triangle to show it is a Table Calculation.

4. Choose Compute Using > Table (down). You should now have a table that shows the rank per subcategory within each region:

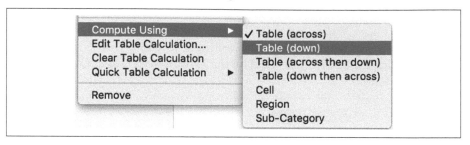

Discussion

Tableau provides a lot of table calculations. Understanding the Compute Using options depends on the data that is in your view. In the solution just shown, the Table Calculation defaulted to Table (across), which means by region, whereas we wanted to calculate the rank using the subcategory Table (down). However, when using Table (down) or Table (across), moving either of the fields, in Rows or Columns, the Table Calculation will recalculate based on the new layout in the view.

I recommend that once you know what you are computing the calculation by, you should change the calculation from the default settings. To do this, go into the Edit Table Calculation option and choose the Specific Dimensions option. This means if you change the layout of the view, the Rank calculation will still compute using Sub-Category, as long as Sub-Category is still in the view.

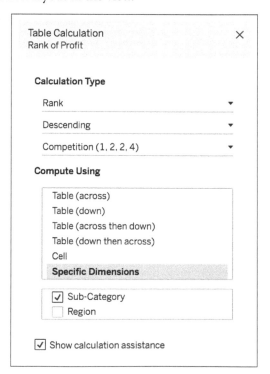

If you remove Sub-Category from the view, you will get a red field, indicating that something has happened with the Table Calculation:

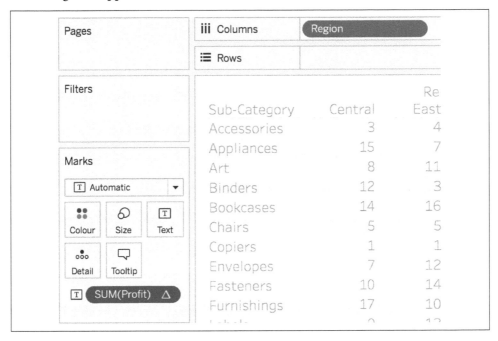

Tableau has four rank options, which change how it ranks the values. We can use the values 100, 95, 95, and 85 to demonstrate the differences:

- Competition rank would give the two 95 values the same higher rank of 2 and give 85 the rank of 4.
- Modified Competition rank would give the two 95 values the same lower rank value of 3, and 85 the rank of 4.
- Dense rank would give the two 95 values the higher rank but would give 85 the next rank value of 3.
- Unique rank would give each value a new rank depending on how the data is sorted, but by default it is alphabetically.

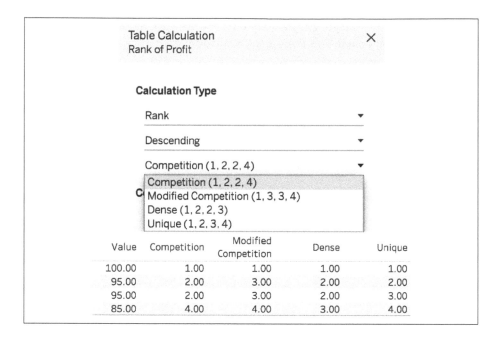

You can also create a rank calculation by using the calculated field option, where you can see the description for the calculation:

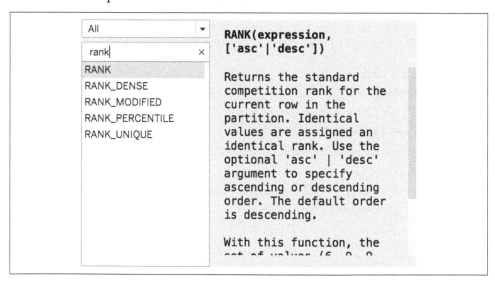

When creating a Table Calculation through a calculated field, you will notice "Default Table Calculation" at the bottom right:

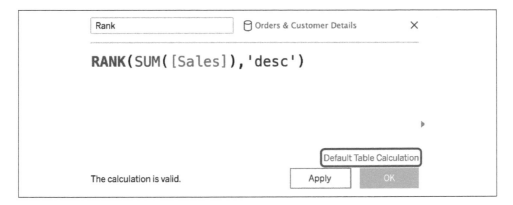

This allows you to specify how you want your Table Calculation to compute. When you click this option, it brings up the Calculation Definition:

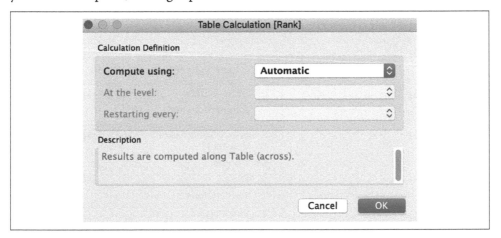

This Calculation Definition allows you to set your Compute Using options. However, when you change the default, if you don't have that field in the view when using the calculation, the Rank Calculation field will error.

For more information about Table Calculations, see Chapter 16.

4.5 Big Actual Numbers

Big Actual Numbers (BANs) are used on dashboards (see Chapter 7) to give the viewer high-level, most-critical information immediately. BANs are usually key performance indicators (KPIs) for the business.

Problem

You want to have a BAN for the latest year of sales so far.

Solution

1. Drag Sales to Text on the Marks card:

2. Drag Order Date to Filters and select Years:

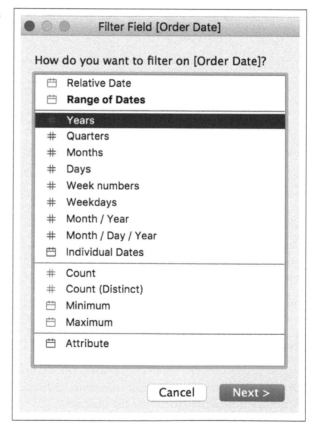

3. Check the box at the bottom, "Filter to latest date value when workbook is opened":

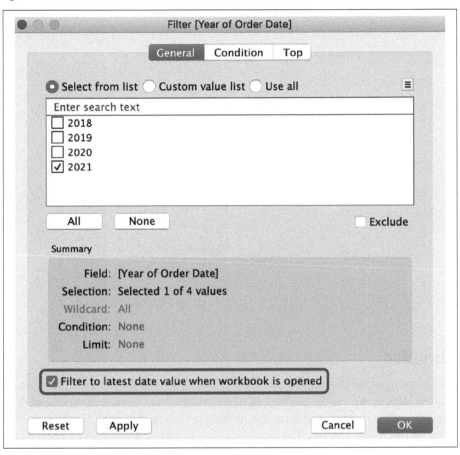

This then shows the sales value for the latest year of data:

$733,215

If you click Text on the Marks card, you can change the alignment to Center, which allows the BAN to be in the center:

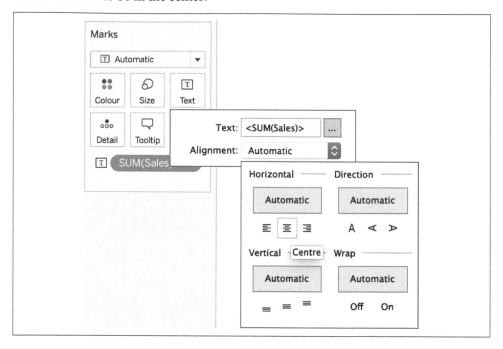

Discussion

BANs are used for the high-level, most critical KPIs on a dashboard. They can be formatted depending on the dashboard. You can increase the size by using Size on the Marks card or increase the font size in the Text properties. You can also change the color of the BAN, which can be done in one of three ways:

1. Use Color on the Marks card and use the default colors or a color picker.
2. Use a color that is in the color palettes. Create a calculated field with any text or letter—this is just a dummy calculation for the color and then add that calculation to Color on the Marks card.

3. Change the color of the font inside the Text properties. Click Text on the Marks card and select the three dots. You can then highlight the text and change the color:

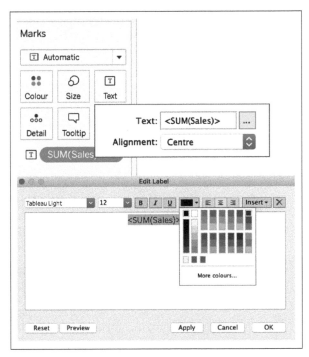

4.6 Calculating Percent Difference

When using BANs, you might want to include a percent change from another value (a previous value or time period, for example). This section will show you how to calculate the percent difference, and Recipe 4.7 will show you how to keep only one value.

Problem

You want to see the percent difference for sales from the previous month.

Solution

1. Filter to the latest year (see Recipe 4.5).

2. Right-click (Option on Mac) and drag Order Date to the Rows shelf and choose MONTH. Then add Sales to Text on the Marks card:

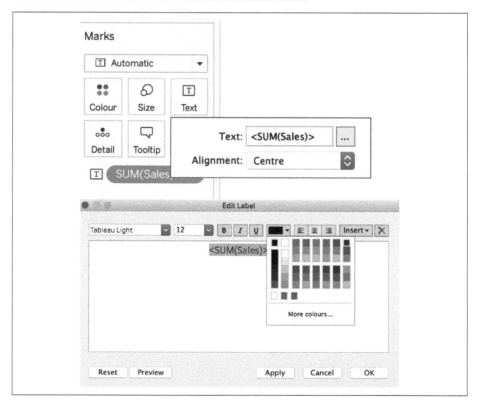

3. The goal is to calculate the percent difference from the previous month in the view. Right-click SUM(Sales); then choose Quick Table Calculation > Percent Difference:

 Always double-check how your Table Calculation is being computed. Also, I highly recommend changing from the default to Specific Dimensions when you are happy with the compute.

This now shows the percent difference for sales compared to the previous month:

Month of O..	
January	0.00% 0
February	-53.83% -23,670
March	190.00% 38,571
April	-37.96% -22,351
May	21.19% 7,740
June	19.70% 8,721
July	-14.57% -7,717
August	39.45% 17,856
September	39.20% 24,746
October	-11.48% -10,090
November	52.29% 40,671
December	-29.23% -34,619

Discussion

The percent difference calculation has the same Compute Using options as previously mentioned in Recipe 4.4. The default this time is Table (down), which is computing by month, because month is the only field in the view. Ensure the Specific Dimensions option is selected. January is blank because it doesn't have any previous values to compare to (i.e., the year filter has removed the option to compare against December of the previous year). See Recipe 4.7.

Currently the default option compares the values to the previous value; however, we can change this option under "Relative to." This means you can select which month you want to compare to. Currently, the default is Previous (month), but you can select Next to look at a future month, First to look at the first month (January), or Last to show the last month (December).

If you want to show the actual difference between the months, add another SUM(Sales) to Text on the Marks card, and then right-click and select Quick Table Calculation > Difference. You

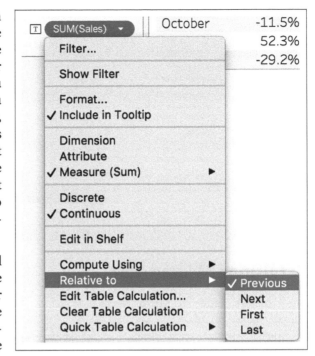

will need to change the "Relative to" option if you have changed the percent difference from the default.

If you drag the new sales field, with the triangle on, to the left toward the Data pane, Tableau will save the Table Calculation as a calculated field. You can then reuse the same calculation across your worksheets. You might need to change the Compute Using option when you drag it into a new sheet.

If you press Ctrl (Command on Mac) and click to drag a field to another part of the view (i.e., the Filters shelf), the Table Calculation will also travel with the field.

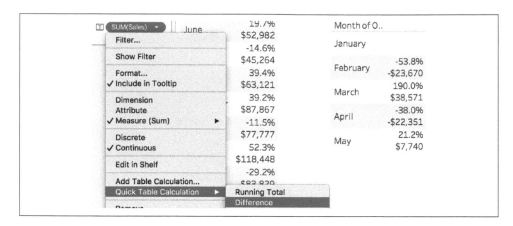

In the next section, we will finalize the BAN by showing only the last row of data.

4.7 Using LAST and Hide

In Tableau's order of operations, table calculations are computed after every filter, meaning that if you filter a year or a month, you will lose the percent difference calculation, like the following:

This recipe uses a table calculation called LAST that can be used to hide data.

Problem

You want to see only the LAST month as a BAN while also keeping the percent difference from the previous month.

Solution

1. Re-create or duplicate Recipe 4.6.

2. Create a new calculated field and type **LAST()=0**:

3. Drag this calculation to the Rows shelf. Notice that you now see True or False. December = True because that is the last value using the default Compute Using of Table (down).

4. Right-click the word False and click Hide:

This then keeps just the original percent difference and difference from previous values:

Last()	Month of O..	
True	December	-29.23% -$34,619

Discussion

In Tableau's order of operations, table calculations are computed last, after all the filters. By using the LAST table calculation and hiding the false results, you can still keep the table calculations for your BANs and show only the latest month in the dashboard.

The LAST table calculation gives each row in your view a number that indicates the distance from 0. In this example, since December is the last value in the table, it would be assigned the value 0. November would be given the value of 1 because you need to go forward one month to get to December.

Month of O..	LAST()
January	11
February	10
March	9
April	8
May	7
June	6
July	5
August	4
September	3
October	2
November	1
December	0

If you want to keep the first record in your view, the opposite of the LAST function is using FIRST, where the same principle applies —the first value gets the number 0. However, the other values start with a negative, indicating the number of marks you have to go backward to get to the first mark.

In Recipe 4.6, I mentioned that January would be blank since it doesn't have a value to compare against. You can use the FIRST function to hide the first value in the list.

Month of O..	LAST()	FIRST()
January	11	0
February	10	-1
March	9	-2
April	8	-3
May	7	-4
June	6	-5
July	5	-6
August	4	-7
September	3	-8
October	2	-9
November	1	-10
December	0	-11

4.8 Custom Number Format

Problem

When computing a difference or percent difference, you want to apply a custom format to the text, parentheses to indicate a negative difference, and an upward or downward triangle for a positive or negative percent difference, respectively.

Solution

1. Start by either duplicating Recipe 4.7 or creating a new sheet with a percent difference calculation.

2. Right-click the Value Difference table calculation and click Format:

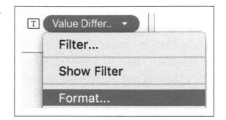

3. Select the Numbers drop-down. Here we can select a variety of options. For the value difference, we want to change it to a currency, have 0 decimal places, and change the negative values to be in parentheses:

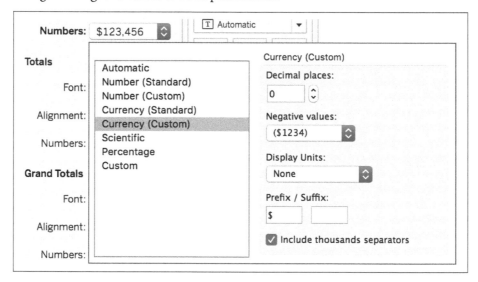

4. Next, you want to change the percentage format. Repeating the same process as before, right-click the Percent Difference Table Calculation.

5. This time in the Numbers format, select Percentage and then select Custom.

6. In this box, you now need to tell Tableau what you want to do with the positives, negatives, and neutrals. Positives and negatives use upward and downward arrows. Use the following text: **▲0.0%;▼0.0%;0.0%**.

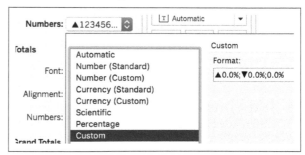

This is used as an alternative method if you want to customize the number even further.

Codes for the up and down triangles are Alt + 30 and Alt + 31

Your final view should look like this:

Last()	Month of O..	
True	December	▼29.2% ($34,619)

Discussion

Custom formatting is an extremely useful key concept because you can customize each measure with a different number format. If you change the format of a field in the view, it will impact only that visualization. If you want to use that format on the measure in different views, you will need to change the default format of a measure, which means whenever you use that measure, it will use the same formatting. However, view-level formatting can overwrite the default formats. To do that, you will need to right-click a measure in the Measures section of the Data pane and select Default Properties. Selecting the Number Format gives you the same options as mentioned earlier.

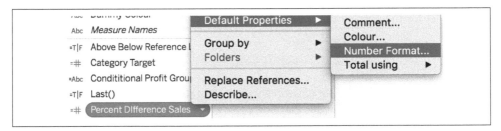

When doing the custom formatting, notice how you use semicolons to separate the figures. This is to differentiate positive values, negative values, and zero values.

4.9 How to Zero Nulls

The zero null (ZN) function can be wrapped around any calculated field to convert nulls to zeros.

Problem

When showing a full year difference or percent difference, you want the first value to show 0 value and 0% instead of null.

Solution

1. Open a value difference calculation (see Recipe 4.6).

2. At the start of the calculation, write **ZN** (and at the end of the whole calculation, close the parenthesis, and then click OK):

```
Value Difference Sales

Results are computed along Table (down).
ZN(
ZN(SUM([Sales])) - LOOKUP(ZN(SUM([Sales])), -1)
)
```

3. Repeat for the percent difference sales:

```
Percent Difference Sales

Results are computed along Table (down).
ZN(
(ZN(SUM([Sales]))
- LOOKUP(ZN(SUM([Sales])), -1)) /
ABS(LOOKUP(ZN(SUM([Sales])), -1))
)|
```

Now the January figures show zero instead of nulls, like the following:

```
                              0.0%
January
                                 0
```

Discussion

The ZN function is especially helpful when you have null values in your data. Having a zero instead of a null is important when creating calculations; you have to be sure that you can treat nulls as zeros before you can use this calculation. The calculation window also describes what the ZN calculation does:

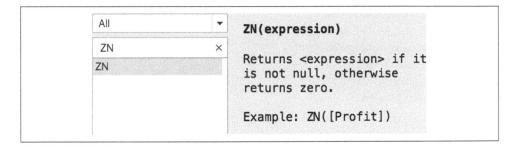

You might also notice that when you converted the table calculations into reusable calculations, Tableau used ZN inside those calculated fields. If any value is a null, the calculation would return a null, which is why it is important to use the ZN function, to convert any nulls to zeros by default.

To use the ZN function, a row of data needs to exist to replace the null with a zero.

4.10 Showing Positive, Negative, or Neutral Values

Problem

You want to color the percent difference based on positive, negative, and neutral values.

Solution

1. Create a new calculated field using the SIGN function and Percent Difference Sales field:

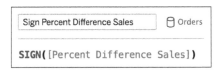

2. Right-click this new calculation and select Convert to Discrete:

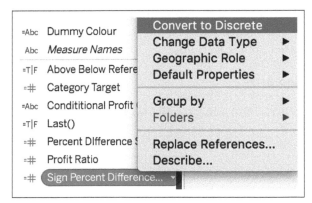

3. Add the calculation to Color on the Marks card and edit the colors. I chose to use orange for negative, gray for neutral, and blue for positive:

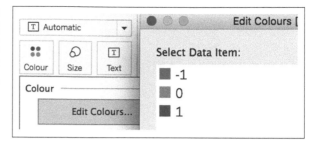

Your final view should look like this:

January	0.0%
February	▼53.8%
March	▲190.0%
April	▼38.0%
May	▲21.2%
June	▲19.7%
July	▼14.6%
August	▲39.4%
September	▲39.2%
October	▼11.5%
November	▲52.3%
December	▼29.2%

Discussion

The SIGN calculation assigns a positive value to the number 1, neutral values to 0, and negative values to –1. Prior to learning this SIGN function, I always used to create conditional IF statements. Using the SIGN function reduces the need to write long IF statements. I learned this from Ann Jackson, a Tableau Zen Master, on one of her speed tipping sessions.

See Also

Ann Jackson and Luke Stanke, *Tableau Strategies* (O'Reilly)

4.11 Calculating a Good, OK, or Bad Status

Using SIGN is useful for positive, neutral, and negative values. But what if you want to see whether a percentage value is good, OK, or bad? These statuses are used in businesses and project management to identify performance or the status of a project.

Problem

You want a status for the percent difference calculations, using Good for greater than 50%, OK for between 0% and 50%, and anything Bad for less than 0%.

Solution

1. Filter to the latest year of data. Create a view with Month of Order date on the Rows shelf and percent difference sales on Text.

2. Create a new calculation and use the following IF statement:

```
IF [Percent Difference Sales]>=0.5 THEN 'Good'
ELSEIF [Percent Difference Sales]>=0 THEN 'OK'
ELSE 'Bad' END
```

3. Add this new calculation to Color on the Marks card and edit the colors to indicate Good, OK, or Bad:

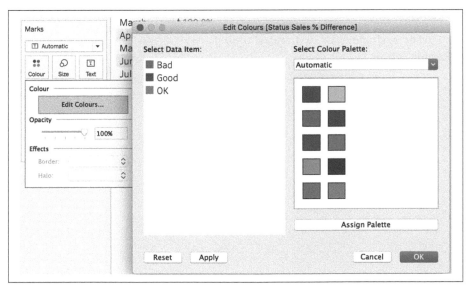

Discussion

These types of statuses are used in project management and are seen as a traffic light system. They are good for visualizing when values are above or below a certain target or percentage. When using this type of status, you might want to be careful with the colors. If people viewing your dashboards are color-blind, they might not be able to see the differences based on the colors. Instead, Tableau has a Color Blind palette that uses shades of orange, blue, and gray:

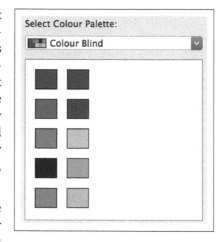

Having this type of status on BANs allows the user to instantly see whether that particular number is above or below the condition that has been set.

See Also

IF statements are mentioned in "Conditional Grouping" on page 111.

4.12 Using Titles as BANs

Problem

You want a single BAN without a tooltip or clickable action.

Solution

1. Right-click (Option-click on Mac) and drag Order ID to Detail; select Count Distinct [CNTD(Order ID)]:

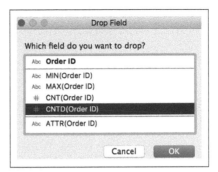

2. Double-click your title at the top. Delete <Sheet Name>, click Insert, and select CNTD(Order ID). This adds the Count Distinct of order IDs to the title.

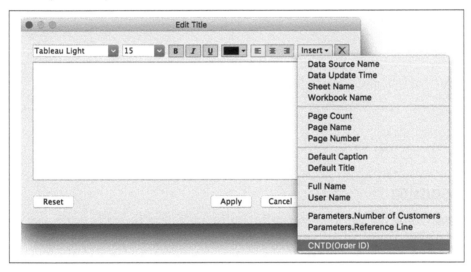

3. Finally, change your Marks card to a Polygon to remove the text in the middle:

Your BAN should now look like this:

Discussion

When using the Title elements as BANs instead of Text, there is a better user experience when you build dashboards in Chapter 7. Titles as BANs are good if you have a single number and don't want conditional formatting with Good, Bad, and OK indicators, and if you don't want or need a tooltip on this specific value. This way also fills an automatic space better without having to use a fake axis calculation to get the space filled correctly. However, if you need to add another dimension, like a Good, Bad, and OK indicator, you will be better using a normal BAN (see Recipe 4.5).

4.13 Using SIZE

Problem

You want to count how many marks are in your view.

Solution

1. Add Order ID to Detail on the Marks card.

2. Create a new calculated field with the following syntax:

3. Add the Size field, just created, to Detail on the Marks card. You will need to compute using Order ID in this case.

4. Add this calculation to the title and change the mark type to Polygon to create a Title as a BAN, as in Recipe 4.12:

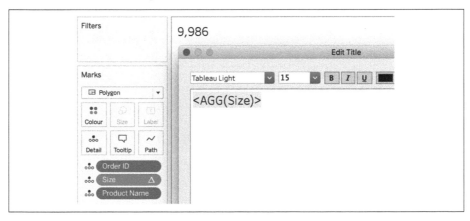

Discussion

Size can be used as an alternative to Count or Count Distinct. The difference is that it counts whatever details you put in your view and how you compute. The Tableau description is as follows:

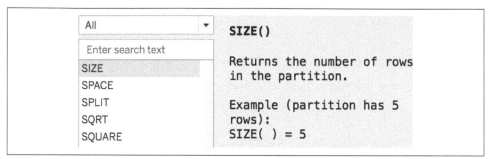

If you added another level of detail to your view, the SIZE value would change. For example, adding the product name changes the title to show 1 to 48:

1 to 48

If we change the Compute Using option to Order ID and Product Name, you'll notice that the title is None:

None

That is because you have changed the Compute Using option, and when you edit the title, it says Missing Field. You will need to re-add the calculation to the title:

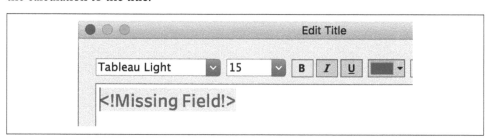

When you do, you'll notice the value has increased to over nine thoousand. That is because there is more than one product name to a single Order ID and Tableau is therefore counting the total number of products within the total number of orders.

4.14 Word Cloud

Word clouds are a visual representation of text in your data, with the option of adding context to those words.

Problem

You want to see the subcategories colored and sized by total sales.

Solution

1. Drag Sub-Category to Text on the Marks card.
2. Change the mark type from Automatic to Text:

3. Add Sales to both the Size and Color properties.

Your final output should look something like the following:

If you press Ctrl, then select and drag a field, you can duplicate this field to any shelf within the workspace.

Discussion

Word clouds have a good visual appeal when looking at text data, and users can quickly see the most frequent words. However, word clouds have some weaknesses. It is difficult to compare the size and color of each word in the view. The layout of the word cloud is also dependent on the word length; if the word has 20 characters, it is naturally going to stand out more than a word with only 5 characters. Word clouds can also have a messy arrangement, and you cannot sort the words by a particular field. The alternatives for this type of chart are a bar chart (see Recipe 3.1) or a tree-map (see Recipe 8.1).

Summary

You can incorporate text into data visualizationsin many ways, by using BANs or labels. Text tables are also a fundamental part of Tableau, especially for those trying to switch from using Excel to creating visuals in Tableau.

CHAPTER 5

Lines

In this chapter, we will focus on using the Line mark type within the Marks card. Line charts are very useful for looking at trends over time. A line chart is mainly used when you are comparing data over time.

Dates can be split into Discrete and Continuous. As mentioned in the introduction, a discrete field will create headers, whereas a continuous field will create an axis. When using Dates, a discrete date will be using an individual date part, which is a stand-alone element like Months or Weeks. When using continuous dates, it will return the date value. When looking at Months, Tableau will return the month and the year instead of just the month.

5.1 Line Charts

Problem

You want to see the sales trend over time, split by year and quarter.

Understanding Dates

When you right-click and drag a date field to any shelf, you will get this pop-up box, which has five sections:

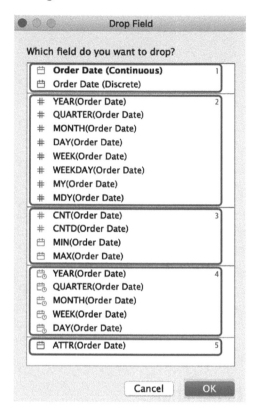

The top section gives you the exact date as either a continuous or discrete field:

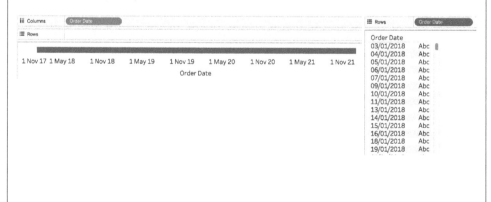

The second section gives you the individual date parts, irrespective of the hierarchy. For example, when using MONTH(Order Date), it will return the individual months without the years:

\# YEAR(Order Date)
\# QUARTER(Order Date)
MONTH(Order Date)
\# DAY(Order Date)
\# WEEK(Order Date)
\# WEEKDAY(Order Date)
\# MY(Order Date)
\# MDY(Order Date)

≣ Rows   ⊞ MONTH(Order Dat..

January	Abc
February	Abc
March	Abc
April	Abc
May	Abc
June	Abc
July	Abc
August	Abc
September	Abc
October	Abc
November	Abc
December	Abc

The third section returns a measure for the dates, which will either count the number of dates, or return a minimum or maximum date.

The fourth section returns the date values, truncated at the level you select. For example, if we choose MONTH, Tableau will truncate the date to the first of a month and contain the specific year:

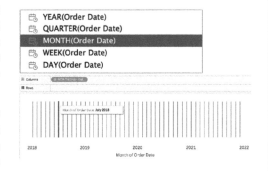

In the bottom section, ATTR stands for Attribute, which returns only a single value if the minimum date equals the maximum date, or an asterisk (*) if the minimum does not equal the maximum.

When using dates, you will need to decide which element you need to use.

Solution

1. Let's start by double-clicking Order Date and dragging Sales to Rows:

2. Use Tableau's automatic date hierarchy and click the plus sign next to YEAR(Order Date) on Columns. Tableau will then drill down to quarters.

Your final view should look like this:

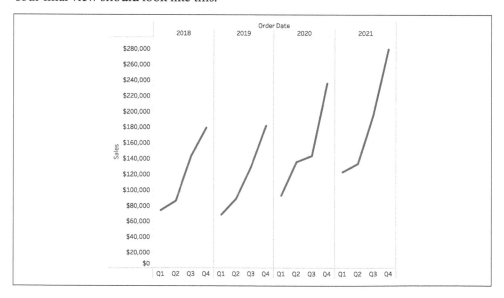

Discussion

Tableau has automatically created a line chart for the data in this view because we are using date fields. Notice that you have two blue fields in Columns. When using multiple discrete fields on either Rows or Columns, Tableau creates sections, which are known as *panes*. In the solution example, we have a pane per year, as that is our first discrete field on Columns.

The problem with this particular view is that it is difficult to compare the values of Q4 in 2018 and 2019 or the seasonal trends. One way to make the comparisons easier is to move YEAR(Order Date) from Columns to Color on the Marks card, like the following:

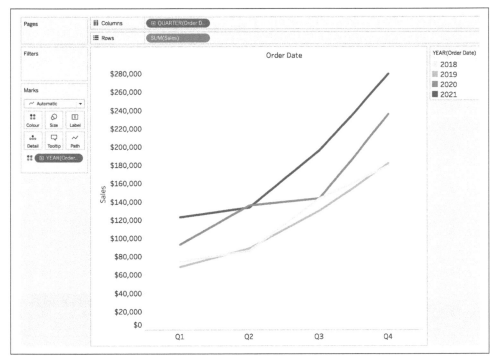

5.2 Continuous Line Charts

Discrete line charts are useful for looking at seasonal trends. *Continuous line charts* show the full range of the trend over time.

Problem

You want to see the total trend of sales by month.

Solution

1. Right-click (Option on Mac) and drag Order Date to Rows.

2. In the fourth section, select the Continuous MONTH(Order Date):

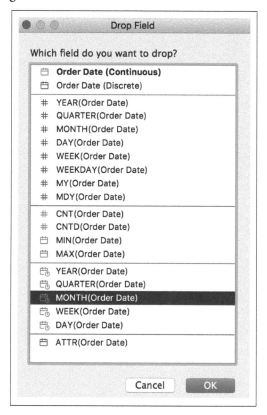

3. Finally, drag Sales to Rows. Your view should look like this:

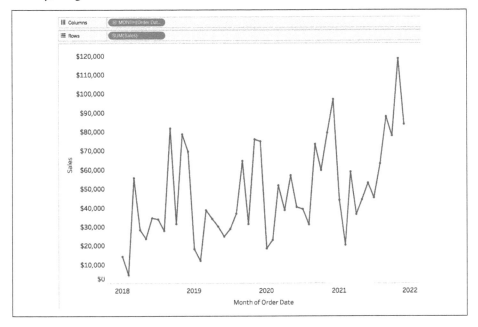

Discussion

Continuous dates create an axis rather than a header. They keep the line connected rather than broken down by the date part, which creates panes. With both sets of line charts, you can add more detail. For example, if you want to see how each category compares over time, add Category to Color.

You can also add markers to the line, which shows where each point on the line is based on the month, by clicking Color on the Marks card:

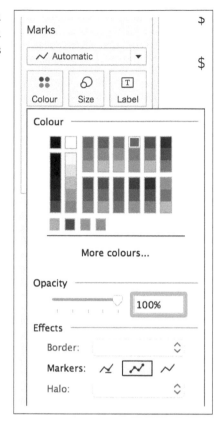

The Marks card has five common properties available. However, when using the Line mark type, you get an extra property called Path. This action allows you to tell Tableau how to draw the line. When selecting the Path property, you have three options: Linear (default), Step, and Jump.

Step (at the top of the following figure) and Jump (at the bottom of the figure) line charts are used to highlight significant changes between the data points:

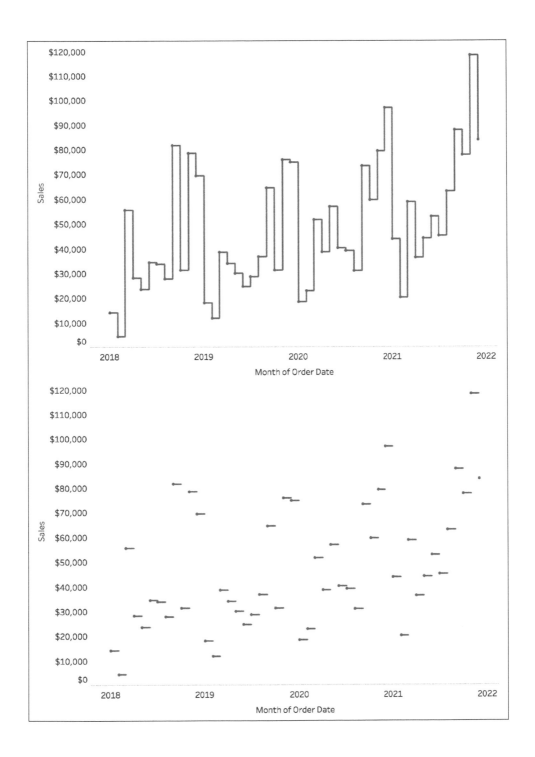

5.3 Trend Lines

Trend lines are reference lines that are a part of the Analytics pane in Tableau.

Problem

You want to see a trend line over time to see if sales are growing.

Solution

1. Start by either duplicating Recipe 5.2 or re-creating it using continuous Months on Columns and Sales on Rows.

2. In the Analytics pane, drag Trend Line onto the Linear option of trend lines:

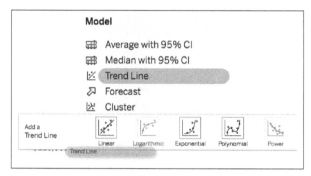

3. Hover over the line to get information about the trend line:

Sales = 29.6555*Month of Order Date + -1.22985e+06
R-Squared: 0.251293
P-value: 0.0002841

Your final view should look like this:

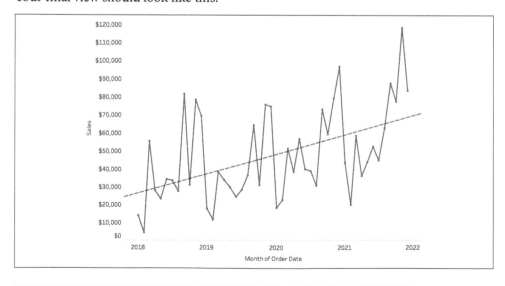

Discussion

A trend line allows you to see the general trend for your data. In this example, we can see that over time sales have increased. The information you get with a trend line shows the equation of the line, the R-squared value, and the p-value.

The *R-squared value* indicates how well the data fits the particular trend line. The closer this value is to 1, the stronger the correlation will be. The *p-value*, also known as the *probability*, indicates whether a model is statistically significant. This particular trend line is significant because the p-value is less than 0.01.

You have two options to get more statistical information about this trend line. If you right-click the trend line and select Describe Trend Line, you will get more information about this trend line:

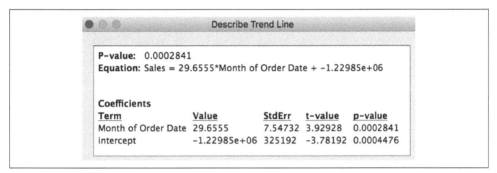

If you right-click the trend line and select Describe Trend Model, you get a detailed description of the trend model:

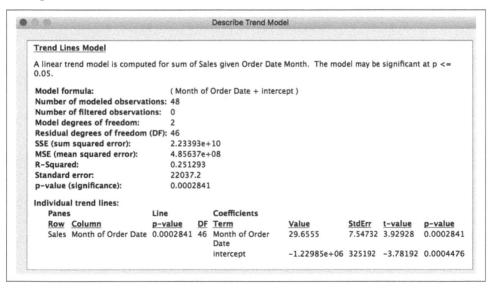

5.4 Forecasting

Trend lines are good at looking at current data, but what if you want to forecast into the future?

Problem

You want to forecast the next 12 months of sales.

Solution

1. Duplicate Recipe 5.2.

2. In the Analytics pane, select Forecast under Model and drag to the view:

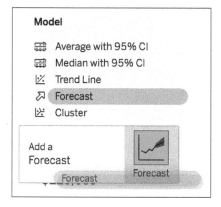

Your view should look like this:

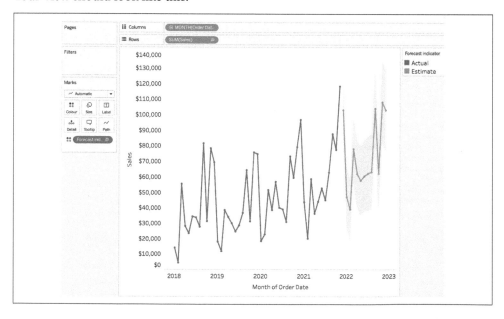

Discussion

Tableau creates a forecast based on the data in the view. To edit the forecast, right-click any point on the line and choose Forecast > Forecast Options. In the Options, you can change how many months you want to forecast and whether to include the last point of data and confidence intervals.

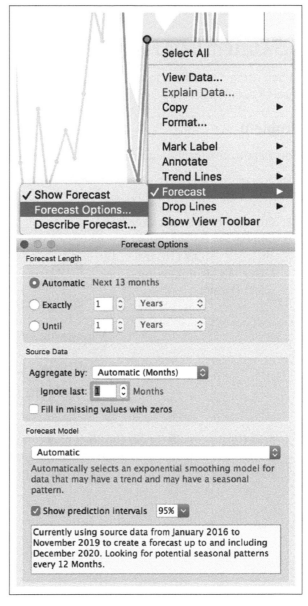

Finally, if you want to remove the line break between the actuals and forecast, remove the forecast indicator from Color on the Marks card.

See Also

The Tableau documentation for the forecast models (*https://oreil.ly/ce6c7*)

5.5 Custom Date Format

When using dates, Tableau will use an automatic date format that is dependent on the level of detail in the view.

Problem

You want to change the automatic date format to a custom date format.

Solution

1. On any recipe that contains a Date Field, Right-click MONTH(Order Date) in Columns and select Format:

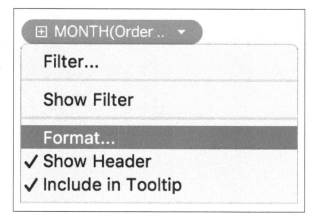

2. Select Dates, under Scale, and then choose one of the options that you want or use the Custom option to specify the format you would like. See the "See Also" at the end of this recipe for full custom date options.

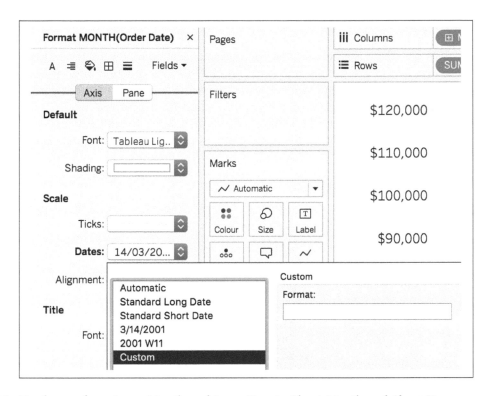

3. To change from Long Month and Long Year to Short Month and Short Year, we need to use the format *mmm yy*.

Your dates should look like this:

Discussion

You can use many combinations of date formats, depending on the level of detail in your view. When you use discrete dates, you will also get different options to choose from. However, if you find yourself using the same format throughout your entire workbook, you will be better off setting the default format for that field. To do that, you right-click the field in the Data pane and then select Default Properties > Date Format. This action gives you the custom option to type in the combination that you are using:

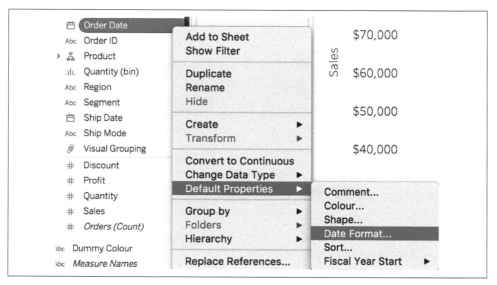

See Also

The Tableau Help Guide (*https://oreil.ly/IqZun*) for custom date formats

5.6 Running Total

At the end of the quarter or year, you might want to look at how your cumulative sales are doing compared to either the previous quarter or the previous year. You'll find this data by using a running total table calculation.

Problem

You want to see how your monthly cumulative sales compare by year.

Solution

1. Right-click and drag Order Date to Columns and select discrete months. Add YEAR(Order Date) to Color and finally, add Sales to Rows. Your starting view should look like this:

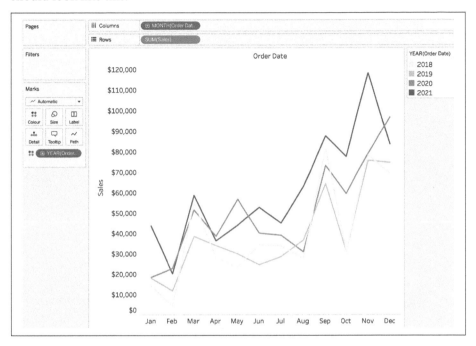

2. To add the running total, right-click SUM(Sales) and choose Quick Table Calculation > Running Total:

3. Because of how we set up the view, Tableau has automatically computed the information by the year (order date). To ensure this is correct, right-click SUM(Sales) again and choose Edit Table Calculation, and then select by Specific Dimensions. As previously mentioned, this will allow the Table Calculation to remain the same if you change the view:

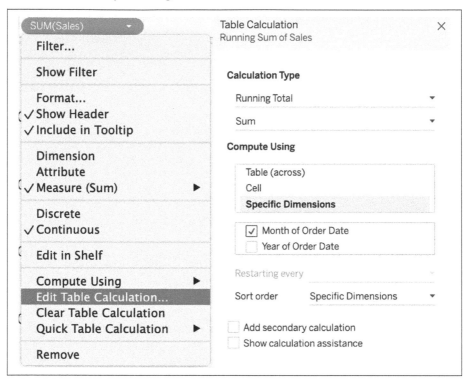

Your final view should look like this:

Discussion

The running total in the example uses SUM, but it doesn't always have to use it. Inside the Edit Table Calculation option, a drop-down allows you to select three other options: Average, Min, and Max. Running average is the average of the current and all previous values. Running minimum and maximum is where all the values are replaced with the lowest or highest value in the partition.

5.7 Year over Year Growth

As with the running total calculation (Recipe 5.6), you might want to compare the percentage growth year over year.

Problem

You want to look at the percentage growth compared to a previous year.

Solution

1. Create a line chart using discrete months on Columns, YEAR(Order Date) on the Color property on the Marks card, and finally, add SUM(Sales) to Rows.

2. Right-click SUM(Sales) and choose Quick Table Calculation > Year Over Year Growth:

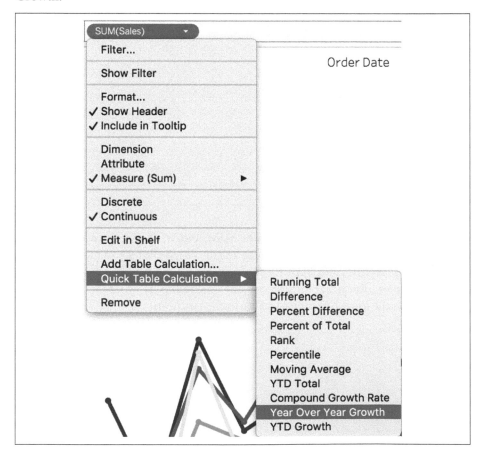

Here is what your view should now look like:

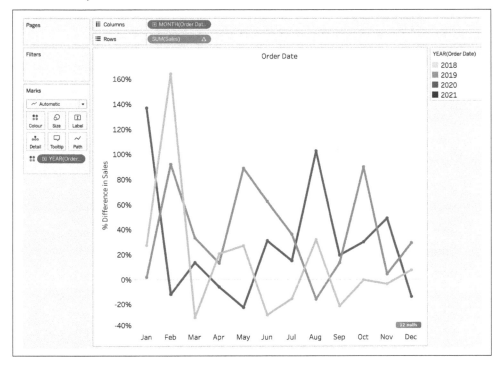

Discussion

The default options for this table calculation set the year-over-year comparison to the previous year—in this case, 2018. That is why an indicator shows 12 nulls—because there is no data for 2018 to compare against. This view compares how each month has grown or declined by year. If you right-click and edit the Table Calculation, you will notice the top drop-down shows just the percent difference. That's because this calculation is just a percent difference with the default set by year. The year over year (YOY) growth option is available only if you have the correct fields in your view (i.e., a year field).

If you used continuous months in your view and didn't have a discrete year field in the view, you can replicate the calculation by selecting Percent Difference instead. This will then compare the previous month year to the current month —i.e., February 2021 will be compared to January 2021 (not February 2020)—which will look like this:

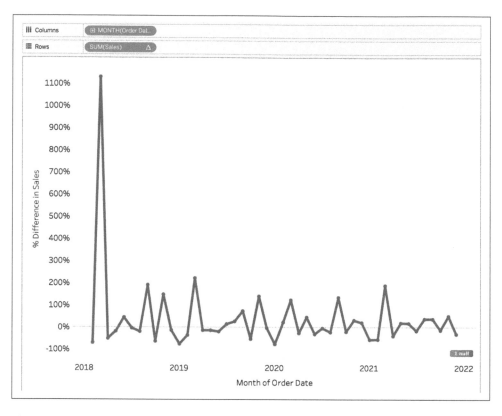

If you want to still compare February 2021 to February 2020, you can edit the calculation. Once you have saved the Table Calculation by dragging it to the Data pane on the left, you can right-click and edit. Inside this calculation, you have two LOOKUP functions:

YoY Calculation   🗄 Orders & Customer Details

Results are computed along Table (across).
```
(ZN(SUM([Sales])) - LOOKUP(ZN(SUM([Sales])), -1)) / ABS(LOOKUP(ZN(SUM([Sales])), -1))|
```

The LOOKUP function uses the specified number as the offset to bring back the previous value. If you change the offset to –12, the function will look at 12 months prior instead of 1 month prior:

YoY Calculation   🗄 Orders & Customer Details

Results are computed along Table (across).
```
(ZN(SUM([Sales])) - LOOKUP(ZN(SUM([Sales])), -12)) / ABS(LOOKUP(ZN(SUM([Sales])), -12))
```

The calculation would make your view look like this:

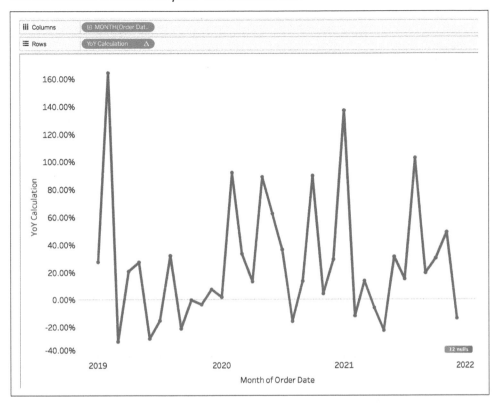

See Also

Recipe 4.6

5.8 Year to Date (YTD) Growth

Both running total and percent difference calculations are powerful. But what happens if you combine the two?

Problem

You want to see the percent difference between each year based on the running total value.

Solution

1. Start with discrete months on Columns, add year of order date to Color, and add Sales to Rows.

2. Right-click SUM(Sales) and choose Quick Table Calculation > YTD Growth:

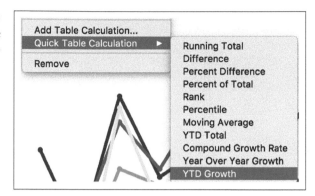

Your final view should look like this:

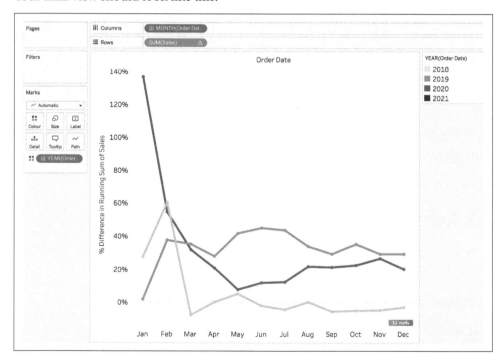

Discussion

The YTD growth calculation is available only when a discrete year field is in the view. If the option is grayed out, you can re-create this using the secondary Table Calculation option. Starting with a Running Sum Table Calculation, you can right-click to edit the Table Calculation. At the bottom of this window, you have the option to "Add secondary calculation." If you select this, you can add a Percent Difference From calculation:

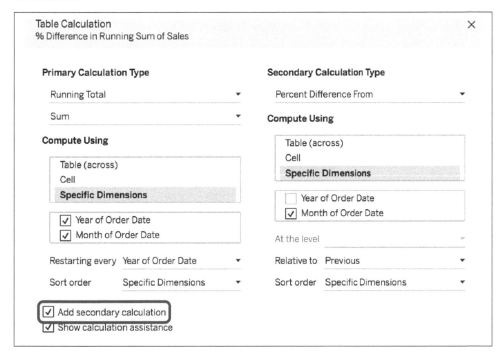

The "Add secondary calculation" will be available only when your primary Table Calculation is either a Running Total or Moving Average.

5.9 Moving Average

Another value you might be interested in is a rolling calculation, or in Tableau's terminology, a moving average. A *moving average* is an average over a subsection of the entire amount of time displayed in the visualization—for example, an average over the previous three months instead of the entire four years.

Problem

You would like to see a six-month rolling average for sales.

Solution

1. Start with the continuous months of order date in the Columns shelf and SUM(Sales) on Rows.

2. Right-click SUM(Sales) and choose Quick Table Calculation > Moving Average:

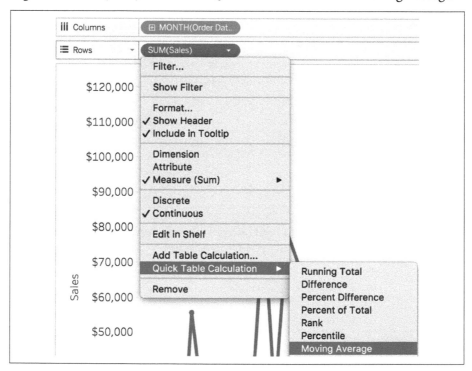

3. When you select Moving Average, the calculation will default to the previous three dates (months in this example). In Tableau, this will be the previous two dates in addition to the date being viewed. This is visible in the Table Calculation window.

4. To change that to six months, right-click the SUM(Sales) field and select Edit Table Calculation.

5. Click the second drop-down. Here, you have the same options as the running total (Sum, Average, Min, and Max). We want to change the previous values to 5 because the current value is selected:

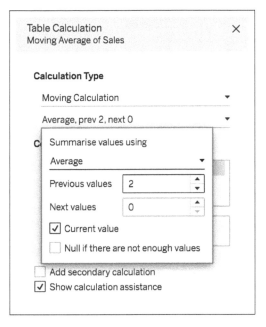

6. Selecting the "Null if there are not enough values" option will remove the first five months of data from the view, because it doesn't have enough values to compute the six-month average, whereas if you keep those values, it will just average the values that it can.

Your final view should look like this:

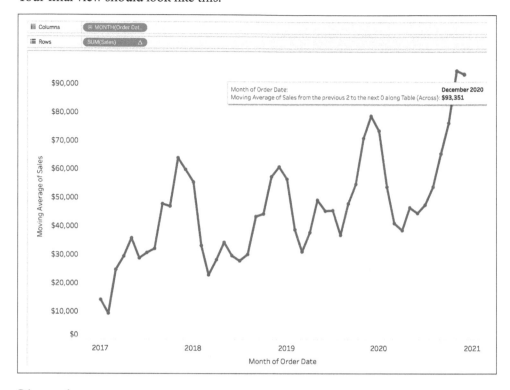

Discussion

A moving calculation can be used to smooth out any fluctuations or seasonality you might have in your data, and it allows you to focus on the long-term trends.

When you are using a moving average rather than the raw values, it might be best to include the phrase "Moving 6-Month Average" in a title or tooltip.

5.10 Slope Chart

Slope charts are used to compare differences from a starting point to an end point. This could be any measure from one year to another.

Problem

You want to compare category sales for the start of the year to end of the year.

Solution

1. Filter to the latest year of data.

2. Build a view that shows Sum of Sales by month, broken down by category on Color.

3. Create a new calculated field using the following: `First()=0 OR Last()=0` (see Recipe 4.7):

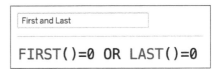

4. Add this calculation to Filters and select True.

Your final view should look like this:

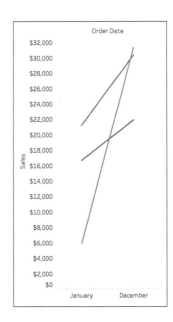

Discussion

Slope charts compare two points in time. This allows you to see the magnitude of growth between the start point and the end point.

Using the FIRST and LAST functions is a dynamic way of creating a slope chart. The easiest way to create a slope chart is to highlight the middle section of the chart:

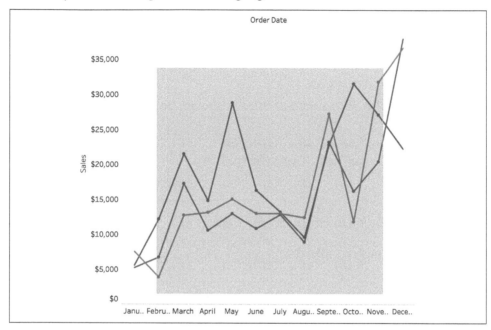

Then when you hover over a single mark from your selection, you can exclude those middle marks:

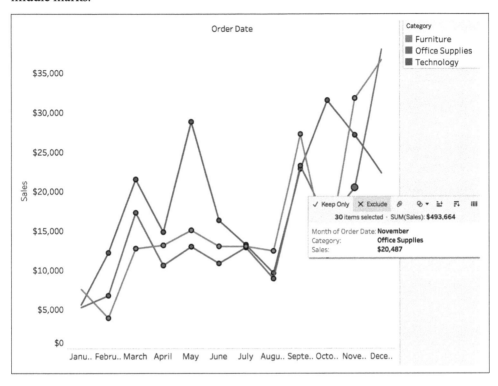

This will give you the same chart as before, except you have an *exclusion* on the Filters shelf instead of the FIRST or LAST calculation we created:

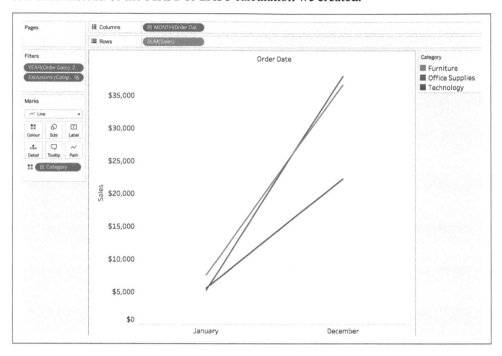

5.11 Sparklines

Sparklines focus on the trend over time and the direction rather than the actual values. Sparklines are used to visualize volatility or outliers. They are usually kept quite narrow on dashboards but still maintain an aspect ratio of 2:3.

Problem

You want to see a sales sparkline for each category.

Solution

1. Add Category and Sales to Rows. Add Continuous Months of Order Date to Columns.

2. When using continuous months, the view automatically fits the width. Instead, we want to make this narrower to make the chart a sparkline.

3. There are many ways to do this, but I recommend choosing Format > Cell Size > Narrower. You can also use the shortcuts next to this option:

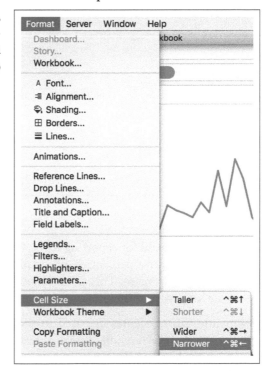

4. When you are using sparklines, you don't need each pane to have the same axis scale, because you are looking at the overall trend of the line rather than the raw values. To change this, right-click the axis and choose Edit Axis. Then select

"Independent axis ranges for each row or column." You should also uncheck the "Include zero" option:

Range

○ Automatic   ☐ Include zero
○ Uniform axis range for all rows or columns
● Independent axis ranges for each row or column
○ Fixed

5. To make sparklines look cleaner, you can make them thinner by reducing the size of the line to the minimum. There is no need to show the axis headers or the date header because you want the focus to be on the trend of the values rather than the raw underlying values.

This is how your final view should look:

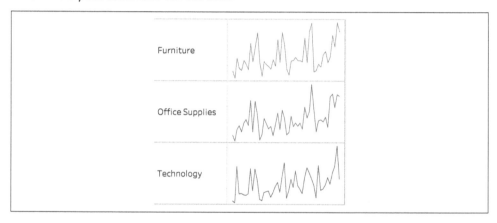

Discussion

Sparklines are useful for reporting KPIs, since they give more context than just using a single number or a percent change, and because of the nature of sparklines—they take up very little space. This example looks at specific categories, but you can easily re-create it by using different measures such as Measure Names and Measure Values.

As mentioned, sparklines are used to show the overall trend and fluctuations in a given measure. However, because of the nature of sparklines, without additional context they could be misleading. I recommend including the last value to allow for a point of reference. To do this, you select Label on the Marks card and check the checkbox. Once the labels are on, you can select the Most Recent to give that additional context:

See Also

Ryan Sleeper, *Practical Tableau* (O'Reilly)

5.12 Small Multiple Line Charts

Sparklines are good for showing the trend of a value, but what if you have a long dimension list, like Sub-Categories? *Small multiple line charts*, also known as *trellis charts*, are perfect for displaying a lot of data in a smaller space.

Problem

You want to show the trend of sales over each Sub-Category by using a trellis.

Solution

1. Create a calculated field called Column Divider and use the following calculation.

Column Divider ✕

Results are computed along Table (across).
```
(index()-1)%(round(sqrt(size())))|
```

Notice that the cursor for this calculation is at the end; it bolds and highlights the corresponding parenthesis. This is a handy feature if you lose track of the number of parentheses you require.

2. Create another calculated field called Row Divider and use the following calculation.

Row Divider
```
int((index()-1)/(round(sqrt(size()))))
```

3. Convert both newly created calculations to Discrete.

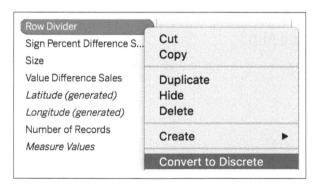

Row Divider
Sign Percent Difference S...
Size
Value Difference Sales
Latitude (generated)
Longitude (generated)
Number of Records
Measure Values

Cut
Copy

Duplicate
Hide
Delete

Create ▶

Convert to Discrete

4. Before we start building the view, and if you are unsure if your data contains a record per Sub-Category per Month Year, then I recommend following this step. You need to create a custom date on your order date. To do this right-click Order Date and then choose Create > Custom date.

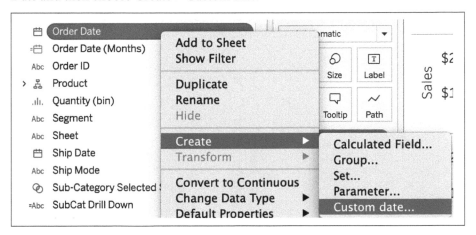

5. In the pop-up box, under Detail, select Months.

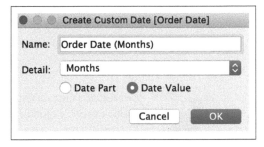

6. Now that we have all of our necessary fields, we can start building the chart. Add Sales to Rows, add the new custom date [Order Date(Month)] to Columns and to Detail. Finally, add Sub-Category to Detail:

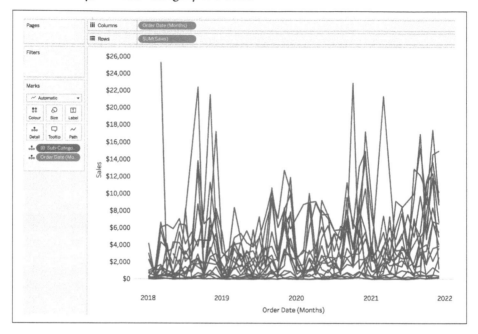

 We created a custom date because not every Sub-Category has a value for every month/year.

7. Add the Column Divider calculation to Columns and add the Row Divider calculation to Rows:

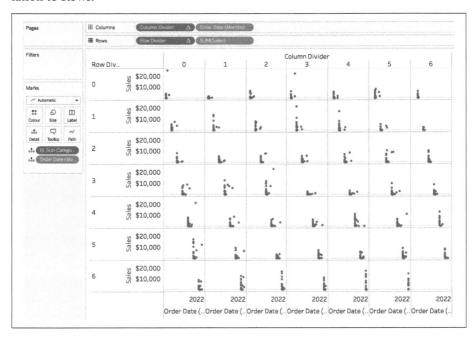

8. To make the small multiple chart work correctly, you need to change the Compute Using option on the Column and Row divider calculations. On the Edit Table Calculation options, make sure Specific Dimensions is selected and Sub-Category is above the custom date. You will also want to change "At the level" to Sub-Category:

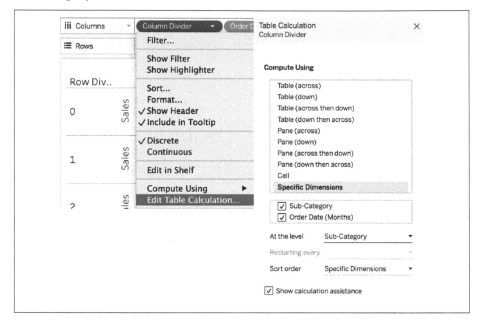

9. Repeat for the Row Divider calculation:

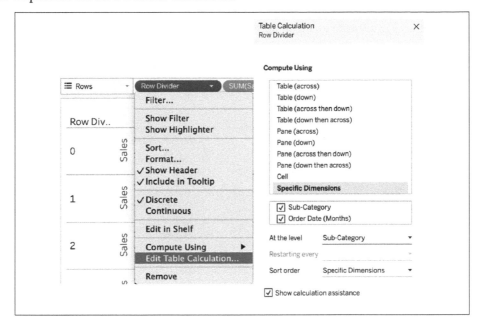

These options will make your view look like this:

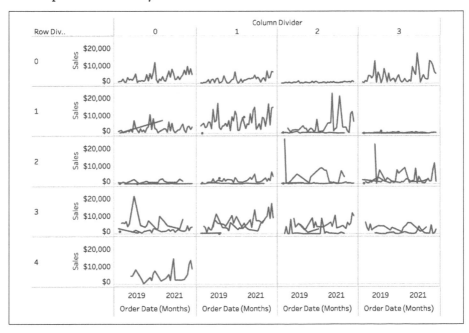

10. You can see that some lines cross over into the different small multiples, which is where there are missing dates. To overcome this, we need to change our custom date on Columns to an attribute:

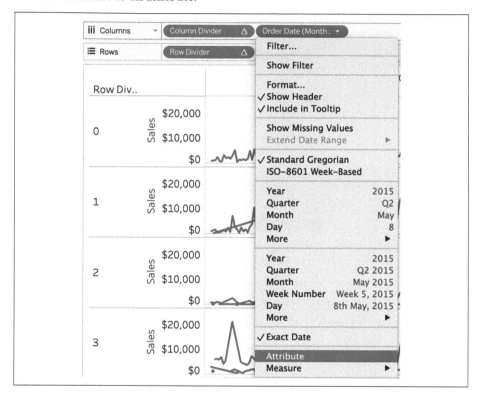

11. Finally, clean up the view. You can hide the headers for Row Divider, Column Divider, and the date field.

Your view should now look like this:

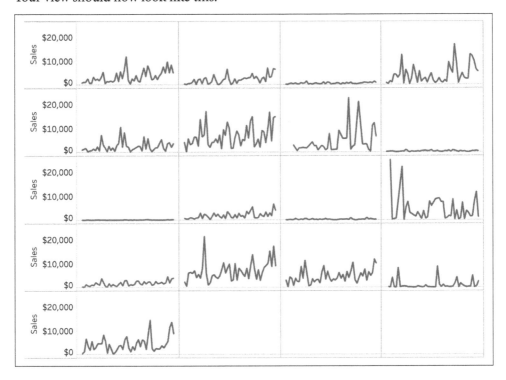

Discussion

Small multiples show a lot of information on one screen. But what does each part of those row and column divider calculations do? These definitions are based on the Compute Using Sub-Categories.

Let's start with INDEX. Index returns the row number of the partition. Index always starts at 1, so if the first partition is Accessories, that will have an index of 1. Using INDEX – 1 allows Tableau to start the count at 0 instead of 1. Both of the calculations use Round, Square root, and Size. Size has previously been discussed in Recipe 4.13. But in this context, it gives the total number of Sub-Categories in the view, which is 17. The next part is the square root. Using the square root allows the number of columns and rows to be dynamically based on the number of squares you want. For example, if we had nine Sub-Categories, it would be ideal to have a three by three grid. This is what the square root tries to do with the data. The final part of the second half of the calculation is Round. This rounds the square root of the size to the nearest whole number; in this case, the final part is 4.

The only difference between the two calculations is that Columns uses Modulo (%) and Rows uses divide (/). The Modulo calculates the number of times a number can fit into another number, and the remainder. In this example, how many times does the value 4 fit into the value of INDEX – 1? So, 0% of 4 = 0 and 1% of 4 = 0, but 4% of 4 = 1, and 5% of 4 also equals 1. This is how we get the Columns value. For Rows, it is the INDEX – 1 divided by 4, and this time the whole calculation is wrapped inside an INT function. Let's use the value of INDEX – 1 is 2. Therefore, 2 divided by 4 equals 0.5, but the INT function rounds the number closest to 0, making this 1. And if we take the INDEX–1 is 7, then 7 divided by 4 equals 1.75—using INT, rounds it down to 1. This is how the Rows calculation works.

 Try to keep your small multiples as simple as possible. If you start adding filters, this will change the way Tableau computes the Table Calculations and may make your chart misaligned.

5.13 Small Multiple Controlled Version

The Rows and Columns calculation uses Size to calculate the number of small multiples. But what if you want to control the number of small multiple squares?

Problem

You want to control the number of squares in a small multiple chart.

Solution

1. Duplicate the Rows and Columns Divider calculations.
2. Create an Integer Parameter. Under Range add a Minimum, Maximum, and Step size for this parameter.

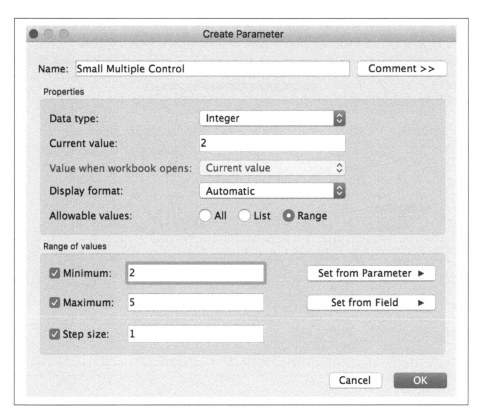

3. In your duplicated Row and Column Dividers, edit the second half of the calculation and replace *(round(sqrt(size())))* after the modulo (%) and divide (/) with the Small Multiple Control parameter.

Your calculations should look like this:

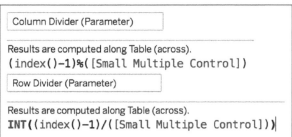

Column Divider (Parameter)

Results are computed along Table (across).
`(index()-1)%([Small Multiple Control])`

Row Divider (Parameter)

Results are computed along Table (across).
`INT((index()-1)/([Small Multiple Control]))`

4. To re-create the view using the new Column and Row Divider calculations, you can duplicate Recipe 5.12 and then replace the previous Row and Column Divider calculations with these new parameter-based calculations.

5. Change Compute Using to Specific Dimensions, at the level of Sub-Category (see Recipe 5.12).

6. The chart will then be controlled by that new parameter. If you show the parameter control and set the value to 3, your chart should look like this:

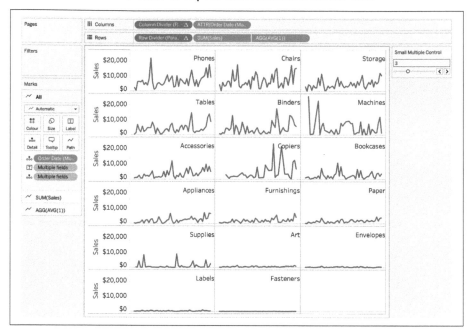

Discussion

Being able to control the number of squares in the small multiple is incredibly powerful, especially if you are designing for different device sizes or have a smaller space on your dashboard.

Here are some handy extra little tips for formatting your small multiple charts. Currently, we don't know what order these Sub-Categories are in. A little trick I like to use is, AVG(1) in Rows as a dual axis. This allows you to add a label above the line chart. Then add Sub-Category to label and select most recent:

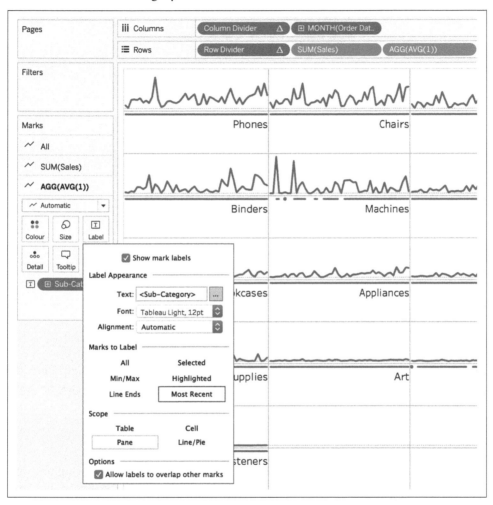

Reduce the opacity of the line to 0% on the Color property of the Marks card. Then right-click the AVG(1) field in Rows and select the Dual Axis option. Finally, remove Measure Names from Detail on the Marks card. You should be left with a view like this:

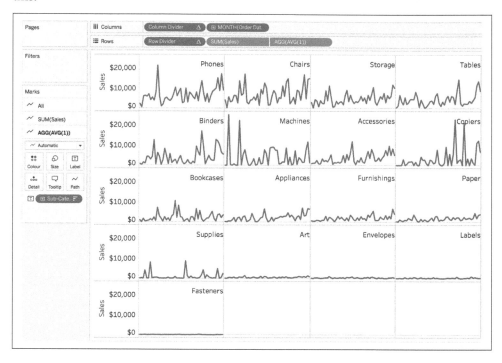

You can also sort the Sub-Categories by the total sum of sales. If you right-click the Sub-Category field on the Marks card, select Sort, by field, and sum of sales descending. This will sort the view.

Summary

The Line mark type is probably the second most used mark type. Throughout this chapter, you have learned the differences between the various date types and how they can be used to build different types of line charts.

You have also learned how to apply some analytical techniques to line charts, like forecasting and trend lines.

CHAPTER 6

Basic Mapping

Maps are a good way to visualize location-based data. Tableau recognizes geographical fields such as country and city, and automatically generates the latitude and longitude. This prevents users from having to manually look up coordinates or pay extra for this type of data.

6.1 Symbol Map

A *symbol map* allows you to show multiple measures on a single map. For example, you can use color and size for two separate measures.

Problem

You want to show every city by profit and sales.

Solution

1. Double-click City from the Orders table. This will automatically create a map.

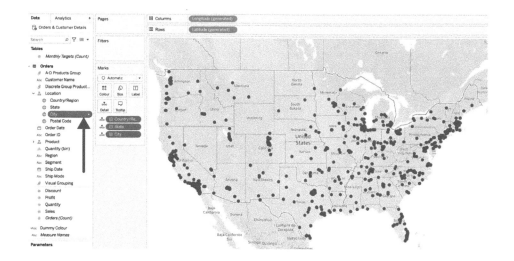

2. Drag Sales to Size (1) and Profit to Color (2):

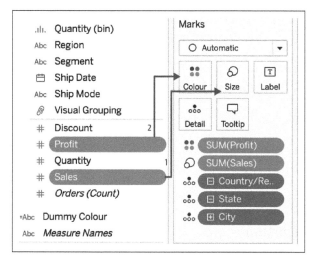

3. You might also want to increase the size of the points. You can do this using Size on the Marks card:

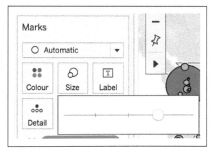

Your final view should look like this:

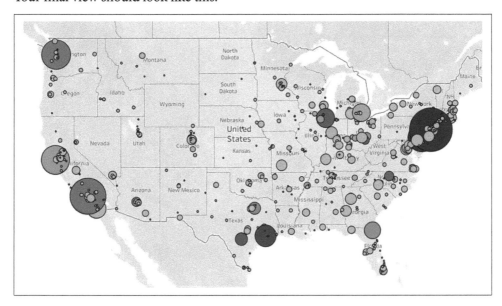

Discussion

Symbol maps are good for showing exact locations—for example, for smaller cities. You should also be careful which measure you use on the Size property. If your measure contains negative numbers, I don't recommend using that field on Size, because you can't accurately display a negative size; therefore, it would be best to use any measure that contains a negative number on Colors.

Editing Sizes

When you have a field on the Size property on the Marks card, you can change the minimum and maximum size of the marks. To do this, click the drop-down on the Size legend and select "Edit sizes":

This will give you a pop-up box that allows you to edit the sizes:

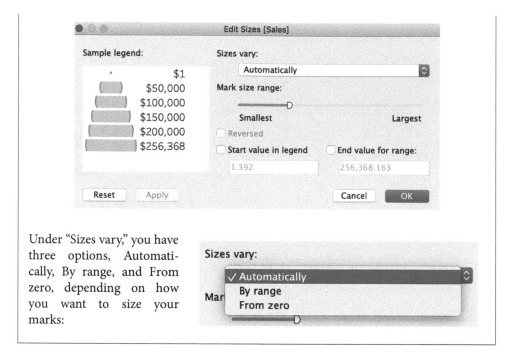

Under "Sizes vary," you have three options, Automatically, By range, and From zero, depending on how you want to size your marks:

Some geographic locations might not be recognized automatically in Tableau. For example, when using the solution, if you remove State from the Marks card, you will get fewer cities and an "unknown" icon in the bottom-right corner of the map:

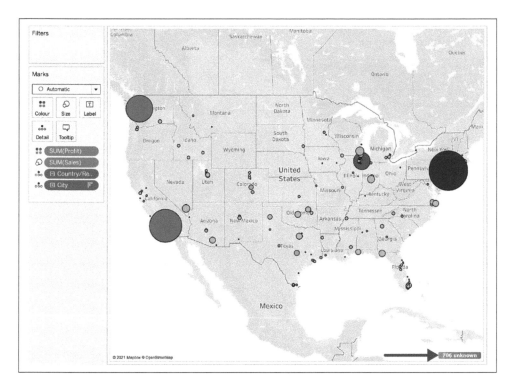

You get the "unknown" icon because many of the city names are ambiguous if they don't have the next level of detail. To edit these locations, you can click the "unknown" icon and select Edit Locations:

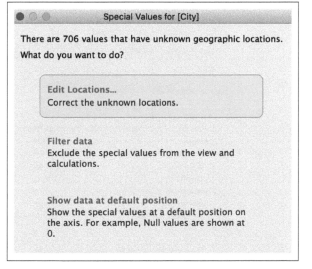

Special Values for [City]

There are 706 values that have unknown geographic locations. What do you want to do?

Edit Locations...
Correct the unknown locations.

Filter data
Exclude the special values from the view and calculations.

Show data at default position
Show the special values at a default position on the axis. For example, Null values are shown at 0.

This will bring up a pop-up box, as in this example, that allows you to pick a State/Province field to map to:

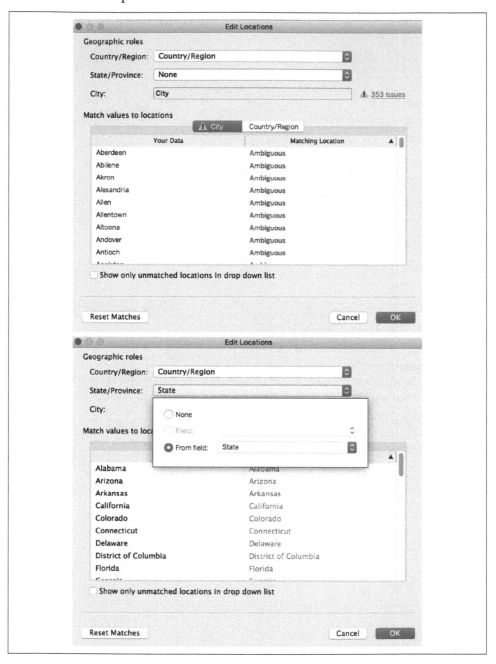

6.2 Filled Map

Symbol maps are good at visualizing the exact location of a city or smaller countries, but a *filled map* can have a bigger impact.

Problem

You want to visualize the profit per state.

Solution

1. Create a new worksheet and double-click the State field under the Orders table:

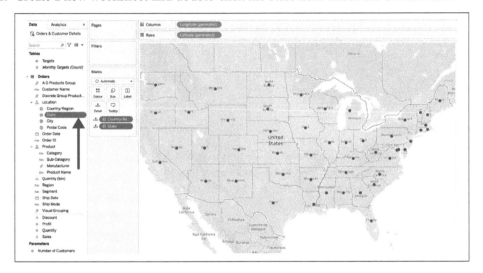

2. Drag Profit to Color on the Marks card:

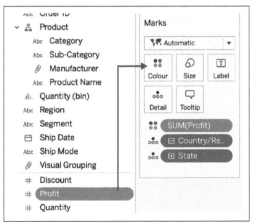

Your final view should look like this:

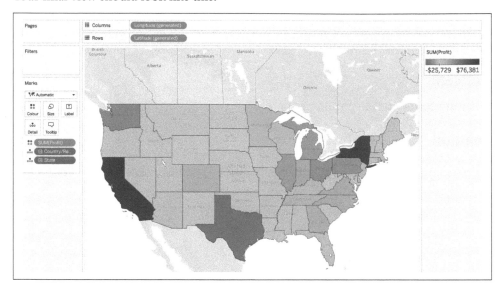

Discussion

This filled map shows the lowest and highest profit by state. If you don't have a filled map, you can select the drop-down under Marks and select the Map mark type:

You can choose from several styles of maps. To change the map type, choose Map > Map Layers:

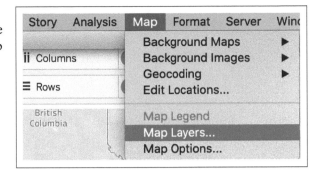

In the Map Layers dialog box, you can change the type of map via the Style drop-down, which has several options. Also, within the Map Layers options, you can change the level of detail on the map, by adding details like terrain and country borders.

Tableau can use the Map mark type only when you have county, state, region, or country geographic fields. It cannot create a filled map at the city or postcode level without bringing in custom geocoding or custom spatial files (see Recipe 18.7).

A filled map should be used for geographic fields of only state or higher. If you want to show cities, you would need to change to a point map.

 If you are visualizing raw numbers (for example, infection rates), it is important that you use a rate that is relative to the population of the area. This is to allow a baseline comparison.

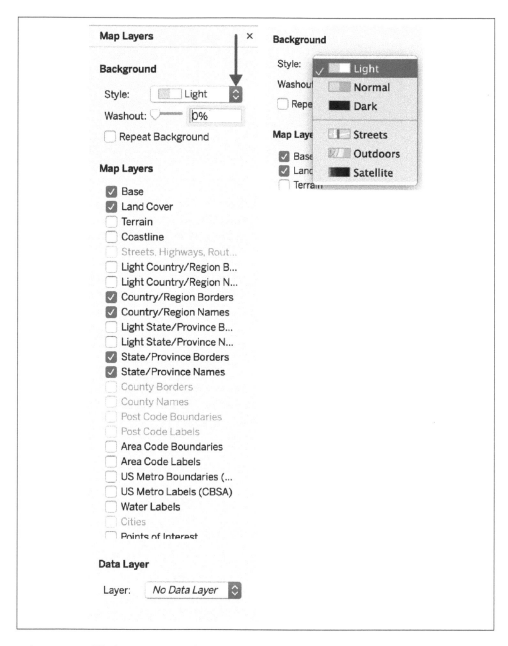

When using filled maps, you should use them with caution. Our eyes are naturally drawn to larger areas on the map. Therefore, when you have smaller countries or states that might have a higher value, their smaller size means you might not instantly see that. This is where a symbol map is better at visualizing the smaller areas.

6.3 Dual-Axis Map

When creating maps in Tableau, you need to use latitude and longitude, whether those generated by Tableau or already in your data set. Both of these fields would be continuous, measures or dimensions, which means they create an axis. As with all other axes in Tableau, you can use them to create a dual-axis chart, which means having two separate Marks shelves. For maps, this means you can visualize at two levels of detail.

Problem

You want to show sales by state, with city by sales and profit layered on top.

Solution

1. Start with a filled state map (see Recipe 6.2) by sales and change the color scale to gray:

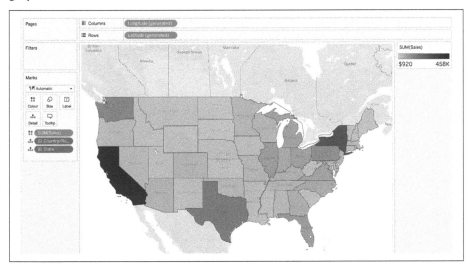

2. In the Rows shelf, duplicate "Latitude (generated)," by pressing Ctrl (Command on Mac) and dragging it next to itself on Rows:

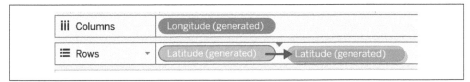

You can also use "Longitude (generated)," duplicating it onto the Columns shelf instead of Rows.

3. On the third Marks shelf on the left, click the plus sign next to State. Because this is a hierarchy, Tableau allows a drill-down to the lower level, which in this case is City:

4. Move Sales to Size on the Marks card and add Profit to Color on the Marks card:

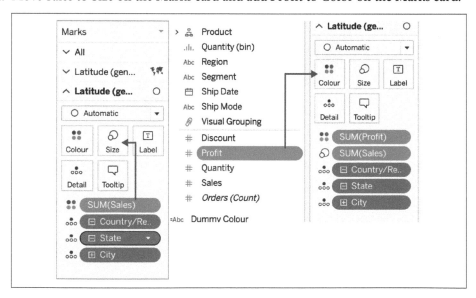

5. The final step is to create the dual axis. You do this by right-clicking the second latitude field in Rows and selecting Dual Axis.

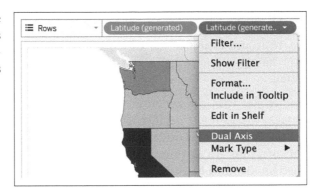

Your final view should look like this:

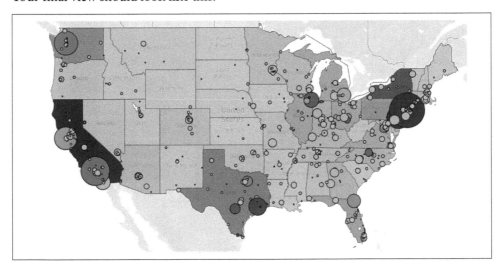

Discussion

As with any dual-axis chart built in Tableau, you have two mark types and two levels of detail.

Sometimes with the symbol map, some of the smaller marks might get lost behind the bigger marks. You should consider sorting the field by the measure. To do this, right-click the City field on the third Marks shelf and click Sort:

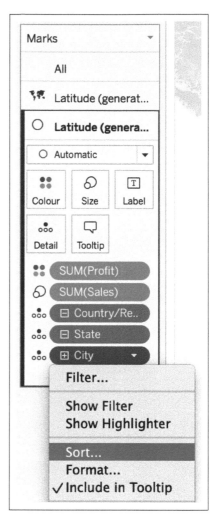

I recommend sorting City by the field you have on the Size property:

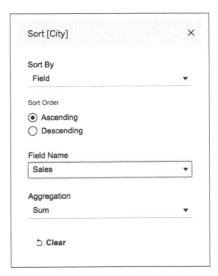

This brings forward all the cities that have smaller sales, which makes your map look like this:

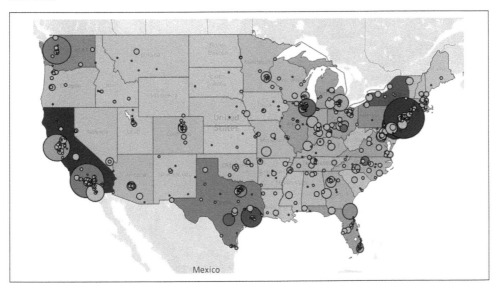

6.4 Mapping Nongeographic Fields

Tableau automatically recognizes various levels of geographic fields. However, region is not one of them. As long as you have other geographic fields, like state or city, in your data set, you can convert a nongeographic field into a geographical one, meaning you will be able to plot it on a map.

Problem

You want to visualize regions on a map.

Solution

1. Right-click the Region field and go down to the Geographic Role. Here you can see all of the standard geographic fields. At the bottom is the option to "Create from." There you can create the geographic field from any of the other geographic fields in the data. For this example, choose State:

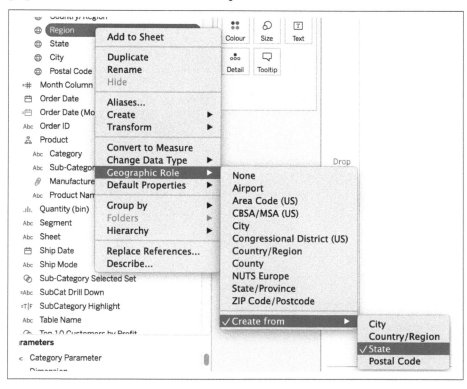

That changes the field to be geographic. You can tell it is a geographic field by the globe next to the field:

2. Now you have your region as a geographic field. Double-click Region, change the default mark type to Map, and add Region to Color. This creates a map of the regions in the data:

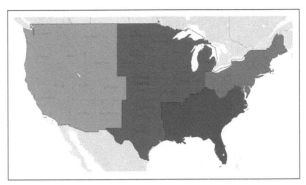

Discussion

Using other geographic locations, like state or country, to enable creation of a new geographic field is a powerful way to be able to visualize nongeographic fields like regions. Having this new geographic field means you can add it into the current location hierarchy. You can do this by right-clicking the Region field and choosing Hierarchy. This option allows you to create a new hierarchy or add to the existing one:

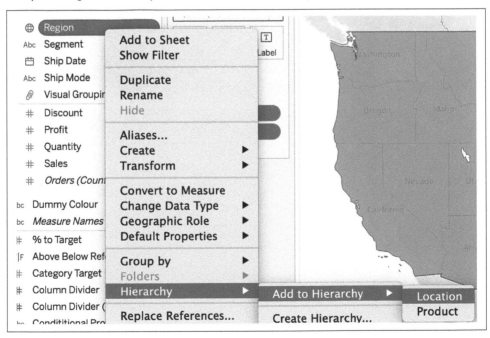

This gets automatically added to the bottom of the current location hierarchy. Region is a level up from State but below Country. You can drag it up to be between those two fields:

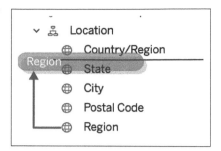

By doing this, you add the drill-down option on the field in the view. Now you can click the plus sign on Region, and it will drill down to the state level:

Creating Hierarchies

You can create hierarchies with any dimension in the data. Apart from right-clicking a field and selecting "Add to Hierarchy," you can also drag and drop a dimension on top of another dimension, and it will automatically create a hierarchy.

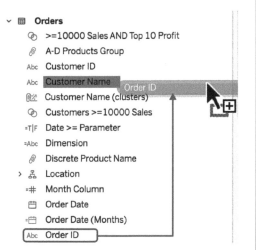

This will give you a pop-up box asking you to name your hierarchy. You have now created a hierarchy by dragging and dropping dimensions on top of each other:

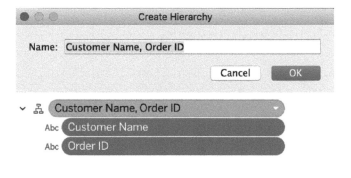

Summary

This chapter has covered some of the basic principles for using geographic fields to visualize data on a map. Chapter 18 covers more advanced recipes for spatial data.

CHAPTER 7

Basic Dashboards

Dashboards are used to bring together the sheets you have created, allowing comparisons between the data simultaneously. Dashboards allow you to add more interactivity among the other sheets you have created. When sharing your work with colleagues, the individual sheets you have created may work well for an analyst, but dashboards really need to help everyone see and understand all aspects of the data in one or several views.

7.1 Build a Basic Dashboard

Problem

You have several sheets that you want to combine to show the overall picture of the data.

Solution

This recipe uses solutions from previous recipes (3.1, 5.2, and 6.2).

1. Create a New Dashboard. The left pane shows all your sheets in the workbook.

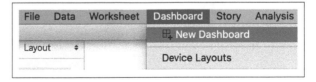

2. Drag the "Map for Dashboard" (Recipe 6.2) onto the dashboard:

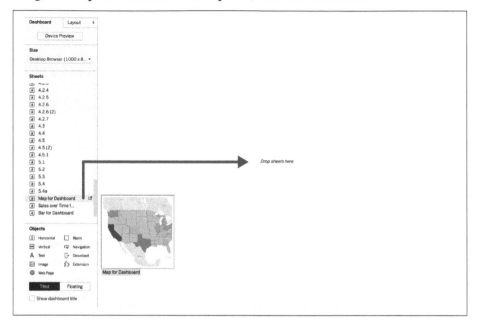

Your view should look like this:

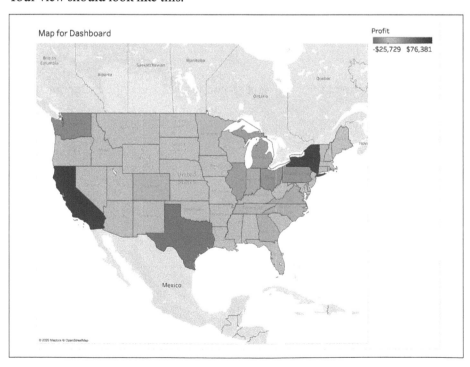

3. Add "Sales over Time" to the dashboard (5.2). This time, we are going to drag it under the map, so it shows the gray shaded area:

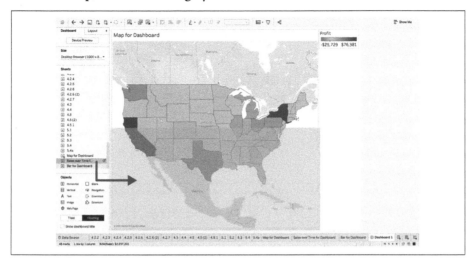

 By default, Tableau creates a tiled layout. For more information, see "Using Horizontal and Vertical Containers" on page 277, "Item Hierarchy" on page 303, and Recipe 19.3.

4. The last sheet we want to add to our dashboard is the Sub-Category Sales (3.1):

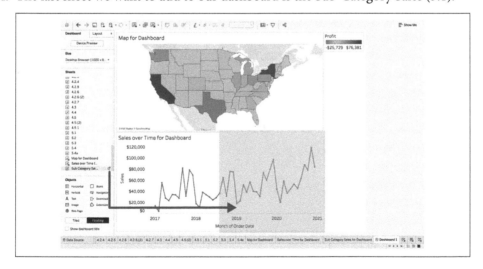

5. When you drag a new sheet onto the dashboard, it automatically shows the worksheet name as the title. In the dashboard, we can edit those to be more meaningful. To edit a sheet title, double-click its title text:

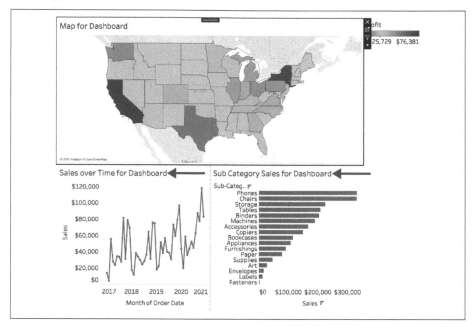

When you double-click a title, the Edit Title box will appear:

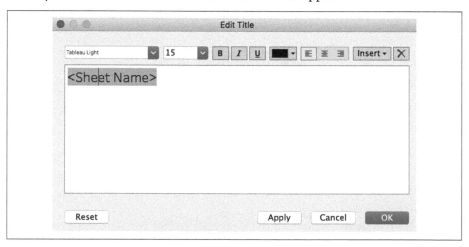

6. Change the title to briefly describe what this sheet is showing. For example, **Profit by State**.

7. Repeat for all sheets on the dashboard.

When you bring a sheet onto a dashboard that has a color legend visible, Tableau automatically brings the color legend onto the dashboard. The legend is currently taking up too much space. I recommend floating this color legend on top of the map.

8. To do this, click the color legend, click the drop-down, and select Floating:

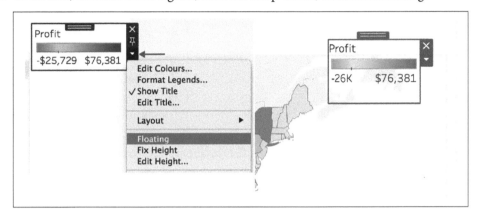

9. You can move the color legend into the whitespace on the map by left-clicking the gray rectangle at the top of the selected object and dragging it to the desired location on the dashboard:

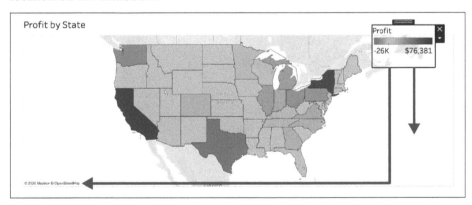

10. Add an overall dashboard title by checking "Show dashboard title" under Objects:

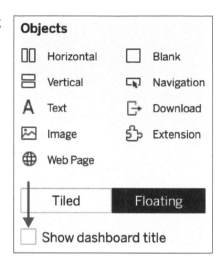

11. To edit this title, double-click it on the dashboard. Or rename the dashboard tab:

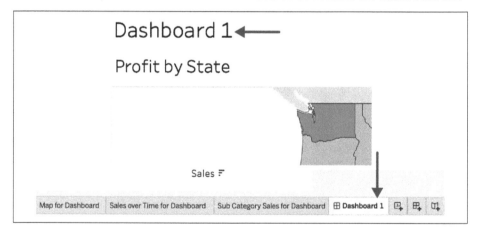

Your final dashboard should look like this:

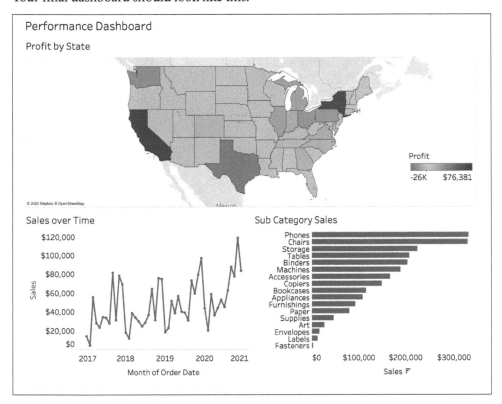

Discussion

Dashboards are a powerful way to bring multiple sheets together in one view—especially when they add the ability to interact with each other.

You can create your dashboards to any size. Tableau has provided some specific sizes of dashboards under the Dashboard tab:

You also have the option to keep the dashboard at a fixed size, or change to Automatic or Range:

The Automatic option will scale up and down depending on the size of your screen. The Range option allows you to specify the smallest and largest size to scale to. Both of these options can produce unpredictable dashboards; what might look good on your screen looks small and cluttered on another. I recommend always keeping a fixed-size dashboard.

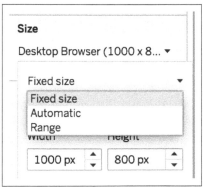

When building dashboards, you might add the wrong worksheet to your dashboard or find that once you have built your dashboard, you want to swap one worksheet for another. You don't have to drag the new sheet in and then remove the old sheet. If you select a sheet that is on the dashboard, which will show a gray box around it, then in the left pane under Sheets, find the sheet you want to replace it with (that isn't already in a dashboard), and click the Swap Sheets button:

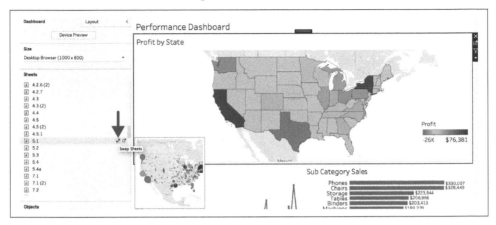

This will maintain any formatting from the sheet on the dashboard, and will replace it with the new sheet, which will make your new dashboard look like this:

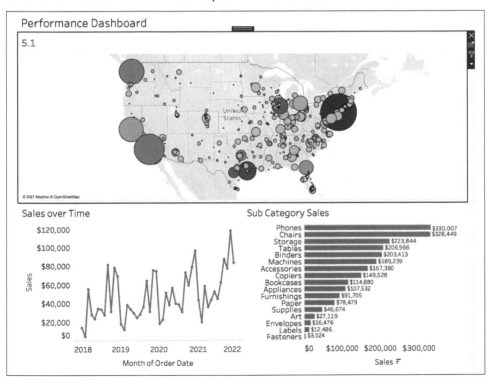

You might notice that when we swapped our sheets, the color legend disappeared, or you deleted the color legend but now you would like to have it on the dashboard. To get the color legend, choose Analysis > Legends. This allows you to add the Color Legend and Size Legend onto the dashboard:

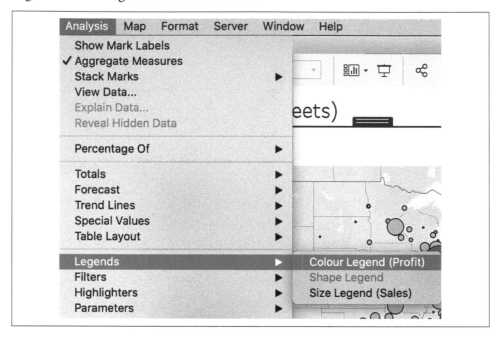

This option also allows you to add any filters or parameters that might also be used on one of the worksheets.

7.2 Adding Basic Dashboard Interactivity

You have the basic dashboard, but now what? To make this dashboard usable, it would be beneficial to add clickable interactivity to allow for further investigation.

Problem

You want to click a state to see how the sales changed over time and how the subcategories have been performing. You also want to click a subcategory, or a time point to filter the rest of the dashboard.

Solution

1. To add a simple dashboard action, select the sheet you want to apply the filter from and select the "Use as Filter" option. You will know it has been selected as the filter will turn white:

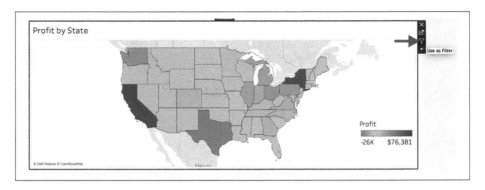

2. This will allow you to select a state to filter the sales over time and subcategories. For example, if you selected Texas, this would happen:

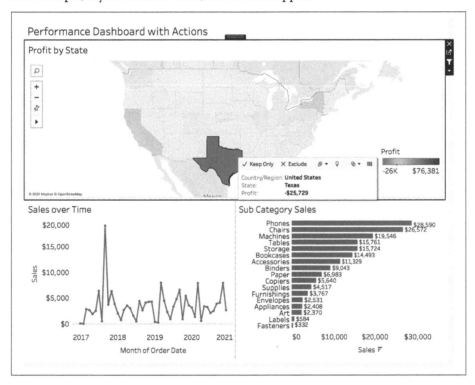

3. You can repeat steps 1 and 2 if you want to add the same filter to the sales over time and subcategory sales. However, this would create three separate actions, whereas for this example, we want to use the same action.

4. To find the action, choose Dashboard > Actions:

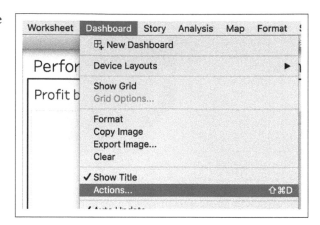

This brings up the dashboard Actions pop-up box:

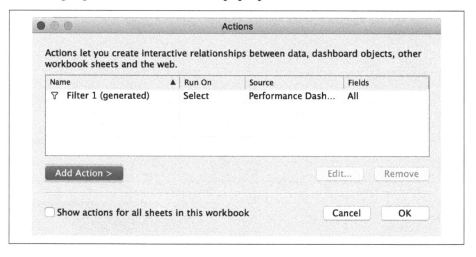

5. This shows you all the actions currently on the dashboard. There are several types of actions. But for now, click "Filter 1 (generated)" and then click Edit:

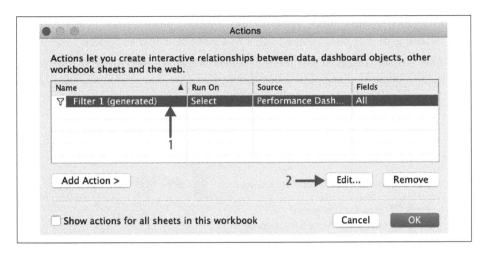

This brings up another pop-up box, which is where we edit the action:

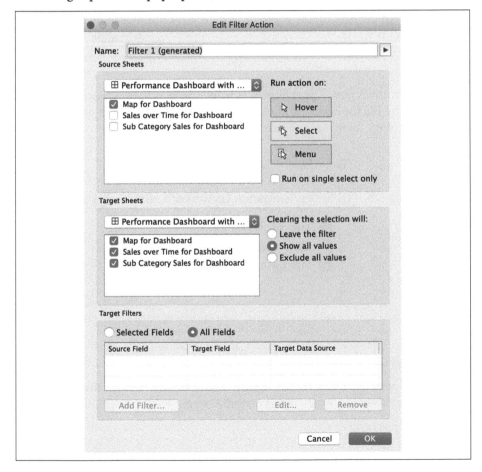

6. First, in these options, I recommend giving the dashboard action a name that is meaningful. For this example, call it **All Sheets Interactivity**.

7. The top section, Source Sheets, is where we select which sheets we want to be able to click to change the other charts. Since we want the same level of interactivity from all sheets, select all three sheets:

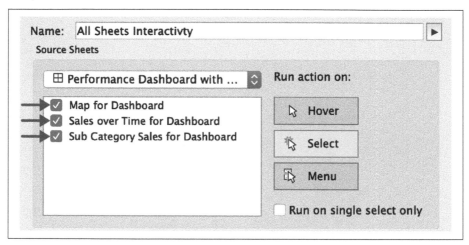

8. The bottom section, Target Sheets, is where we select which sheets we want to be filtered by the top sheets. Once again, we want all sheets to interact with the others; therefore, leave these all selected:

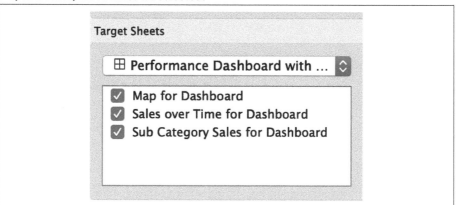

9. Now that we have our filter set up, click OK.

10. Now, if you click any other sheet, it will filter the rest of the dashboard. For example, when you select Tables on Sub-Category Sales, it will look like this:

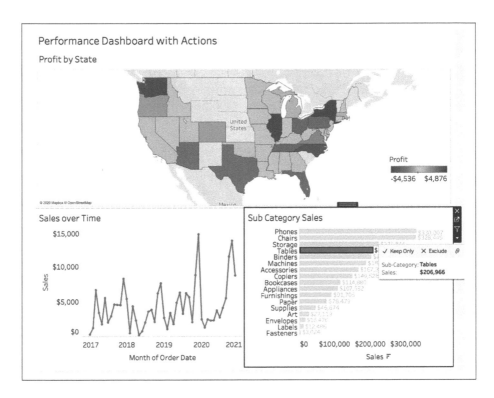

 You can tell this action filter is working because some of the states on the map went null, indicating there are no records for tables in those states.

Discussion

Adding these filter actions allows the user to find their own insights from the dashboard.

Some options have not been mentioned in this recipe. The Edit Filter Action options has a section called "Run action on," which has three options. The option used in this recipe was the default when clicking the filter button on the worksheet, and that is Select. This means when you click a mark, Tableau will automatically filter the dashboard.

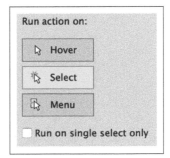

The Hover option means as soon as you hover over a mark, Tableau will filter the rest of the dashboard, but similarly as soon as you hover off a mark, the filter will change again.

The other option is Menu, which means when you click a mark, a hyperlink will appear inside the tooltip, and you have to click the hyperlink to filter the dashboard.

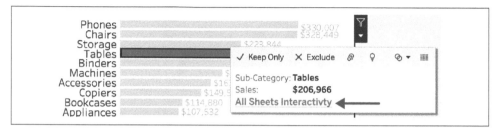

 The default options are "Run action on: Menu" and "Clearing the selection will: Keep filtered values" when you first create an action from scratch.

The other section in Edit Filter Action is "Clearing the selection will." By default, the option is "Show all values." The other two options are "Leave the filter" and "Exclude all values."

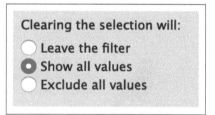

The option "Leave the filter" will keep the target sheets as filtered, so if you select tables, when you deselect or click off tables, Tableau will keep the other sheets on the dashboards filtered to tables.

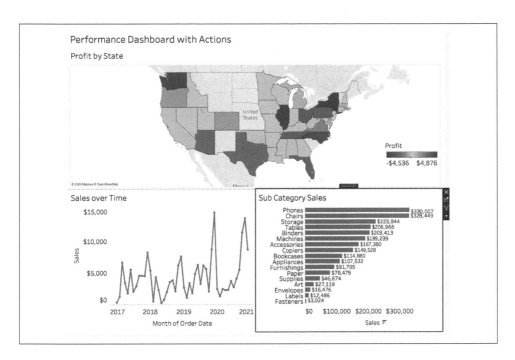

"Exclude all values" will filter out all the values in the other sheets:

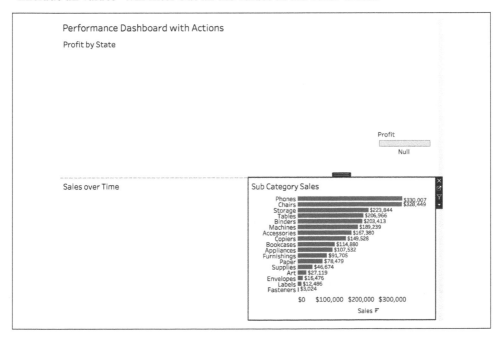

However, using both of these options could mean misleading the user, or thinking the dashboard is broken, if they are not aware of the filter action. It is recommended to add the filter action to titles to make the user aware of the filter applied to the other sheets. And in case screenshots are taken, the reader will be clear on what is included and what isn't.

You also have the ability to select which fields you want to use as part of the action. In the bottom of the action window, you have Target Filters. By default, this is set to All Fields:

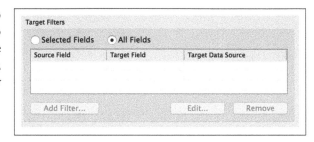

If you change this to Selected Fields, you can choose the fields you want to use. For example, if you want to filter only by category instead of at the subcategory level, you can select Add Filter, and choose Category:

This will allow the action to filter using only the category selected. However, an issue with using the Selected Fields option is that if the field isn't on one of the sheets, you will get a warning to tell you there are missing fields on multiple worksheets:

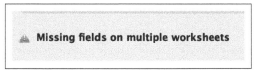

This means when you are selecting your source sheets, you need to make sure that only the sheets that contain Category are selected or you need to add Category to detail on the other worksheets.

7.3 Objects

Dashboard objects are elements you can add to your dashboard that can make building the dashboard easier, enhance a user's experience, and add further functionality. Tableau has nine object types.

Using Horizontal and Vertical Containers

Containers are used to make building the dashboard easier, especially when you want to create fixed-size objects.

Top Tip 3: Follow Tableau Zen Masters and Tableau Ambassadors

Being able to master using containers can really help when you are designing business style dashboards. Containers are very powerful.

Problem

You want to fix the size of the worksheets on a dashboard.

Solution

1. Starting with a new dashboard, drag a Vertical container to the view:

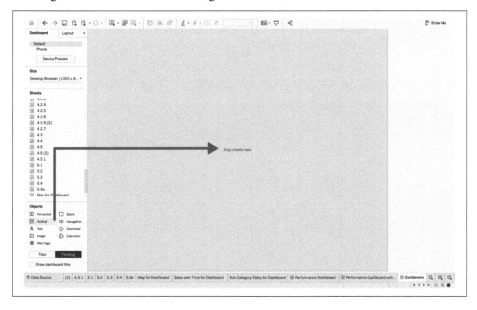

This will leave a blue outline on the dashboard.

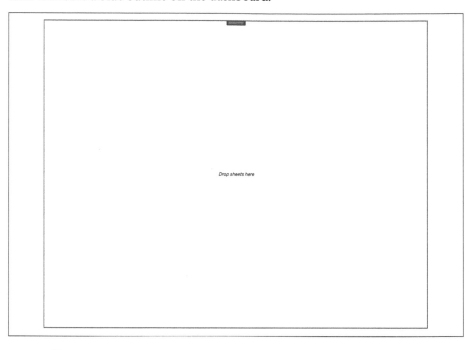

Drop sheets here

Once you have a container, it can be helpful to drag a blank object into the container to allow for easier placement of sheets or objects. Once you are happy with placement, you can then remove the blank from the container.

2. Using the same order of sheets from Recipe 7.1, drag "Map for Dashboard" into the darker blue outlined area:

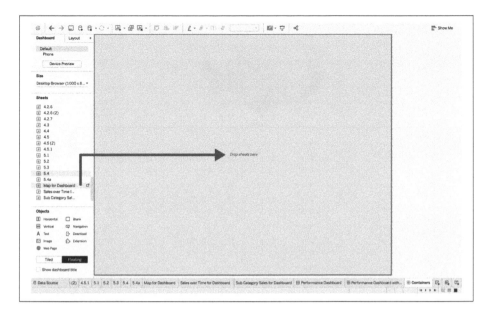

3. This sheet is now inside the container we added first. Next, add in a horizontal container below the map, but still inside the container, which is marked with the darker blue outline. When you drag a horizontal container inside any other container, it will show a narrow gray section where you are dragging it to:

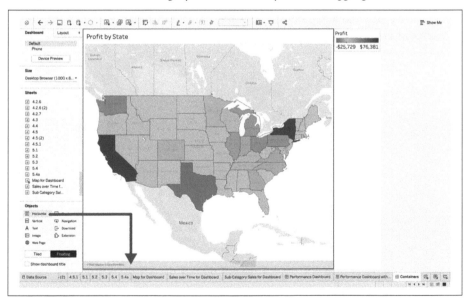

Your view should now look like this:

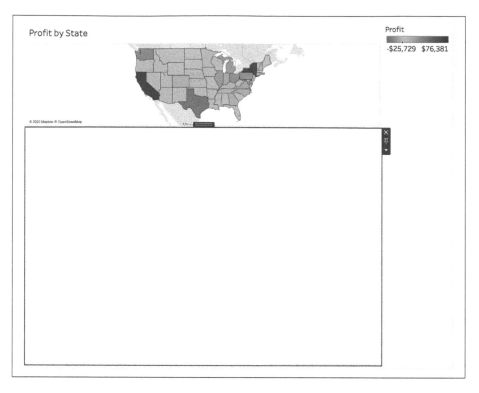

4. Now we need to add "Sales over Time" and "Sub-Category Sales" to the second container we added. You can drag "Sales over Time" into the container:

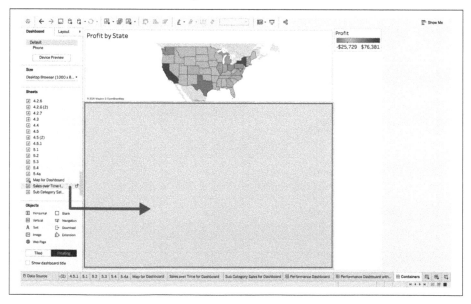

5. Drag "Sub-Category Sales" to the right-hand side of the "Sales over Time" chart. Again, this will drop it into the container as long as the darker blue outline is visible:

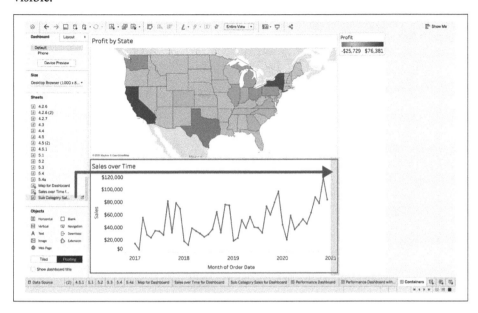

Your final view should now look like this:

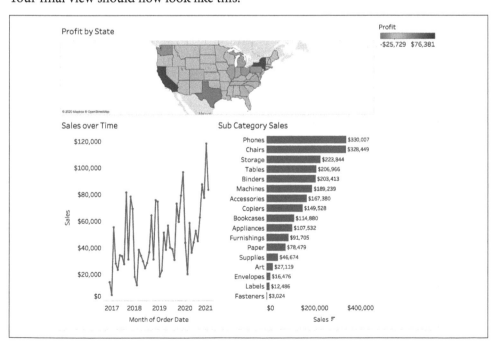

Discussion

The map doesn't fill half of the dashboard because the containers use data in each of the views to determine the space it needs. You can overwrite this default in several ways. The visual way is to find the bottom of the container under the map and drag it down to the required height:

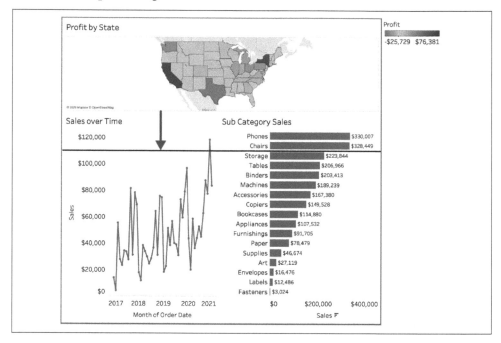

 If you have used a horizontal container, you can fix only the width of a chart; if using a vertical container, you can fix only the height of the chart.

The other way is by using pixels. To change the height of the map, select the sheet, click the drop-down, and select Edit Height:

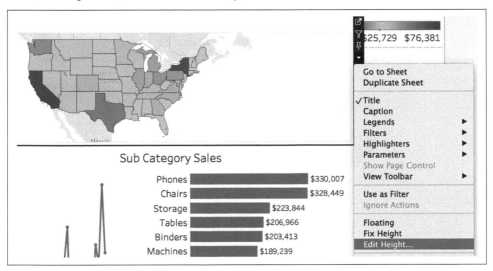

This allows you to specifically set the number of pixels for the chart:

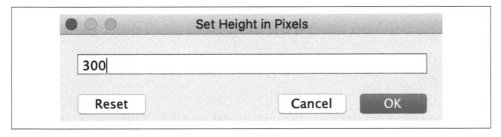

You can also use containers to distribute evenly, that is, to allocate the same amount of space to each chart in the container. This is especially useful if you have three or more sheets inside a container. To do this, you need to first select the container. You can either do this by selecting the sheet and double-clicking the gray lines at the top of the sheet (1), or use the drop-down menu and choose Select Container (2):

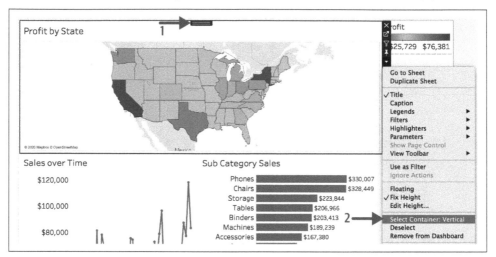

 You can also use the Item Hierarchy to locate a sheet or container (see "Item Hierarchy" on page 303).

This selects the outer container. You can tell the container is selected by the darker blue outline around the sheets. From here, you need to click the drop-down on the container and select Distribute Contents Evenly:

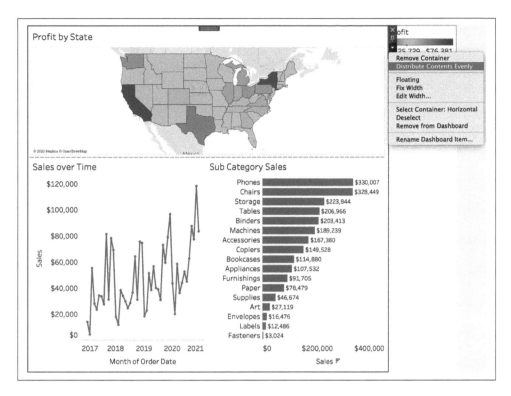

This option will evenly distribute the two sections in this container.

Adding a Text Object

Adding additional text to a dashboard can be useful to add extra context or instructions for the actions.

Problem

You want to add additional text to the dashboard.

Solution

1. Start by duplicating a previous dashboard created in this chapter. You can do this like duplicating a worksheet, except you are right-clicking the dashboard to duplicate.

2. From the Objects section, drag Text onto the dashboard until you see the gray line at the bottom of the dashboard.

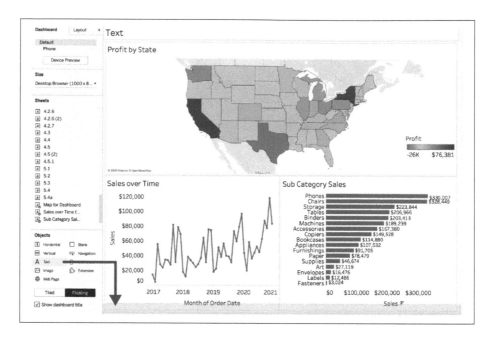

The Edit Text editor automatically pops up.

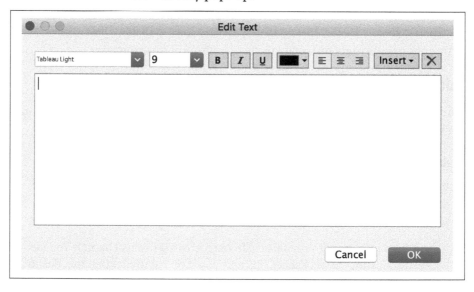

3. In this editor, you can add any text you want on your dashboard.

Your final dashboard should look like this:

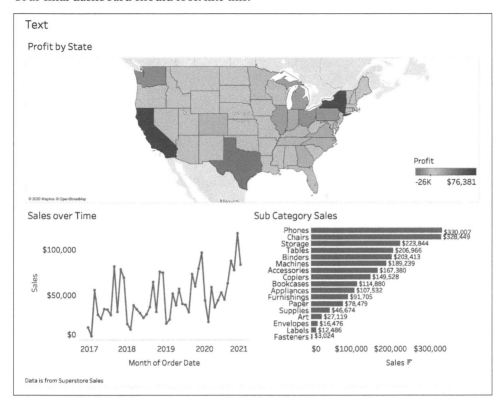

Discussion

Text objects are useful for adding additional text to the dashboard. This could include titles, subtitles, or instructions on how to use the dashboard.

When you use the Text object for titles, be aware that you will not be able to insert fields from any worksheets within your text object; you can add only parameters into these text objects:

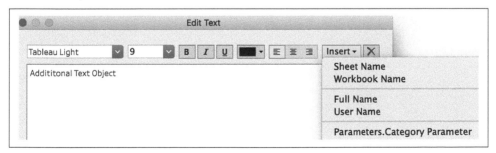

Adding an Image Object

Being able to add images (sparingly) onto the dashboard allows the dashboards to stay within a corporate brand by adding a logo or including a reference image.

Problem

You want to add a company logo to the dashboard.

Solution

1. Duplicate a previous dashboard.
2. Add a Horizontal container to the top of our dashboard:

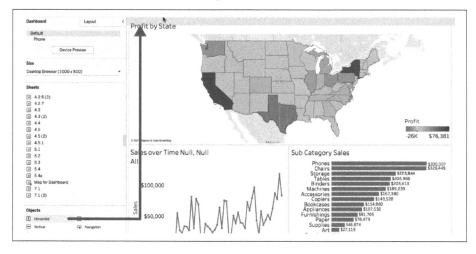

3. Add a Text object to the top of the dashboard above the map. This can act as your title:

4. After editing the text, this text box needs to be resized. Drag the bottom of the text box up to where you are happy:

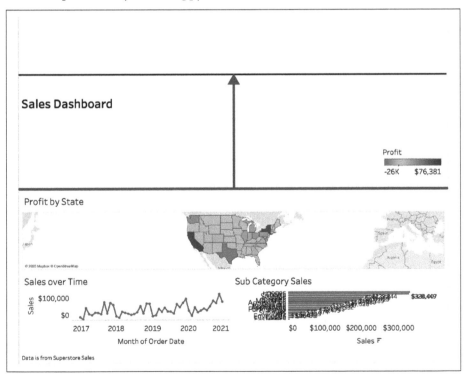

5. Now we are happy with the height of our title. We need to add a logo. From the Objects section, drag the Image option to the top left of the dashboard, inside the horizontal container, next to the title:

When you drop the Image object onto the dashboard, you will be given a pop-up box to edit the image:

6. In this pop-up option, click Choose to select your image from your desktop or documents. You also have the option to Fit the image to the size of the space you have given it on the dashboard, and to center it within that space.

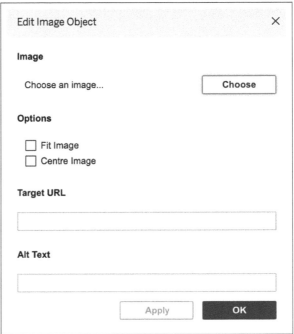

7. When you have selected your image, click OK. Resize until you are once again happy with the size of the logo:

Your final dashboard looks like this:

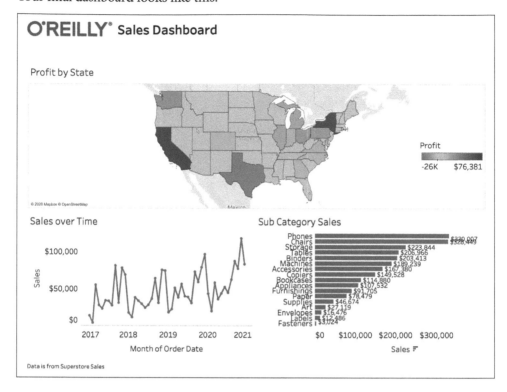

Discussion

Adding images to your dashboard can really help them stay on your corporate brand by adding the company logo. You should use images sparingly throughout the dashboard—while they can make the dashboard more engaging if it fits the audience it is intended for, for others it could make the dashboard look cluttered.

Adding a Blank Object

Blanks are used to give space to sheets on the dashboard. They were especially useful before padding was introduced in Tableau. Padding is discussed later in this chapter. This technique is still used and is useful when picking up older workbooks.

Problem

You want to add a gap between the logo and the title to allow breathing room.

Solution

1. Duplicate a previous dashboard.
2. From Objects, drag Blank between the logo and the title:

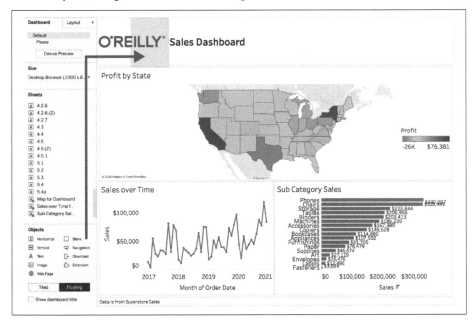

3. Resize the element until you are happy with the size. Your final view should look like this:

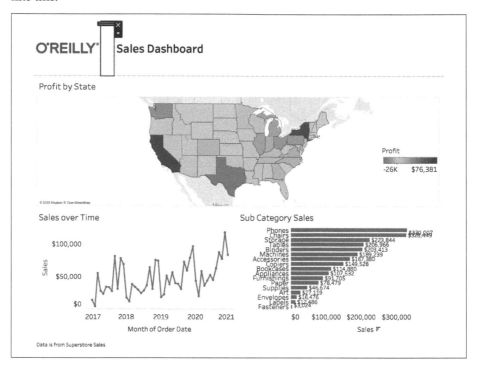

Discussion

Blanks can be used to add whitespace or line breaks to a dashboard. This is particularly important when you have several charts on a dashboard, since it can stop dashboards from looking cluttered. Whitespace is important for letting charts breathe.

For more information about designing dashboards using whitespace, see Ryan Sleeper's *Innovative Tableau* (O'Reilly), Chapter 63.

The remaining objects are discussed in Chapter 19.

7.4 Layout Tab

When working on a dashboard, you will notice two tabs at the top left: Dashboard and Layout. The remaining recipes in this chapter talk about the options inside the Layout tab, which include adding borders and backgrounds to sheets and objects, adding padding to objects, and understanding the "Item hierarchy" in Tableau.

Borders

Problem

You want to add borders to the sheets on your dashboard.

Solution

1. Duplicate a previous dashboard. Select the "Sales over Time" sheet on the dashboard.

2. Under the Layout tab on the left, click the Border drop-down:

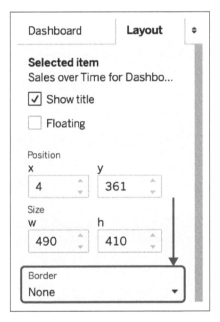

3. In this drop-down, change the style (1), the width (2), and the color (3) of the border:

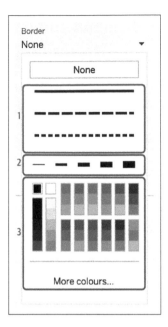

4. Select the first solid line. Your view should look like this:

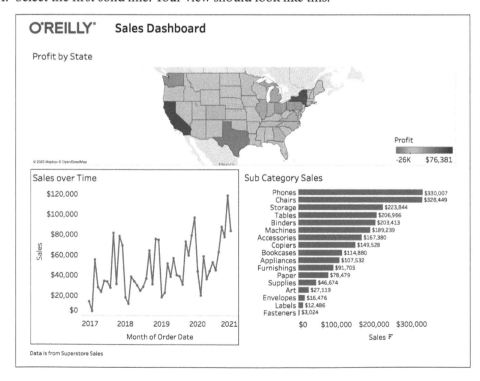

Discussion

You can add borders to sheets or the whole dashboard to separate sections of the dashboard. You can also use borders on containers to help make sure you are adding the sheet into the correct container, especially if you have a dashboard with many containers.

Background

You might want to use different background colors for different sections of the dashboard or for color legends. You can add a background to any element on the dashboard.

Problem

You want to add background colors to the color legend on your dashboard.

Solution

1. Continue with the previous dashboard. Click the color legend.

2. On the Layout tab, click the circle under Background:

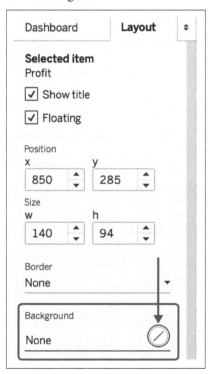

3. Here you can select the color and opacity of the color.

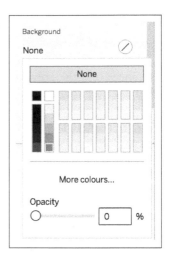

After selecting a light gray, here is what the final view should look like:

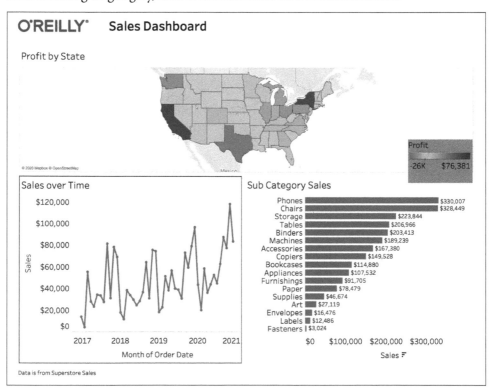

Discussion

Adding background colors allows for segmentation and design of your dashboard.

You can also add a background to the whole dashboard. From the toolbar at the top, select Format > Dashboard.

This brings up a new formatting panel on the left side. By changing the default shading, you change the background color for the whole dashboard:

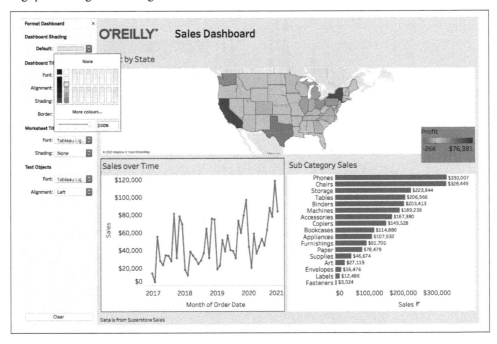

However, as you might notice, this changes the dashboard background, but the sheet backgrounds are still white. This is a default option that cannot be changed for the whole workbook. But you can change it on a sheet-by-sheet basis. Right-click the whitespace on the "Sales over Time" sheet and select Format:

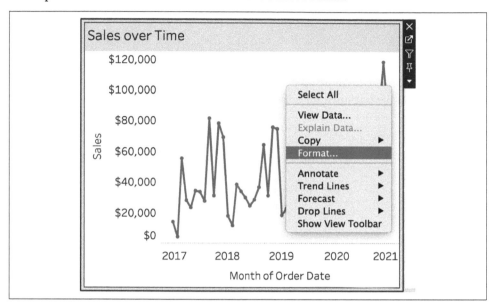

Then you will need to select the paint bucket. In the first option, select None:

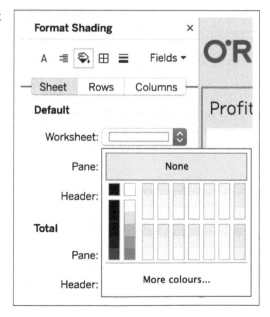

This is what your view will now look like:

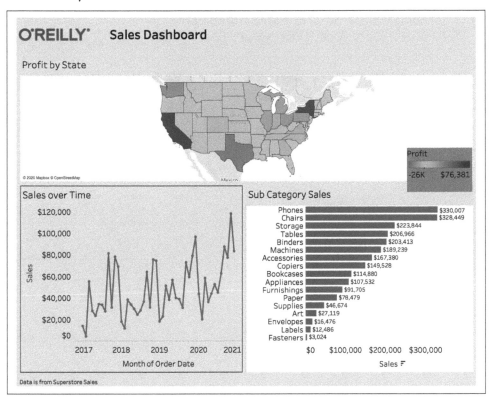

This can be repeated for the other sheets on the dashboard.

Using Padding

On our dashboard, notice that the title on the color legend is very close to the top-left edge. Using padding is a way to ensure there is a little space between the outside and inside. There are two types of padding, inner and outer.

Problem

You want to add padding to the color legend on your dashboard.

Solution

1. Continuing with the previous dashboard, select the color legend and navigate to the Layout tab.

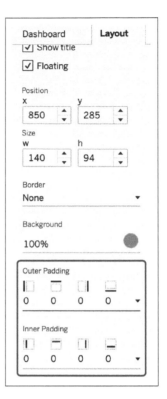

2. Add 5 pixels to both the Inner Padding and Outer Padding options:

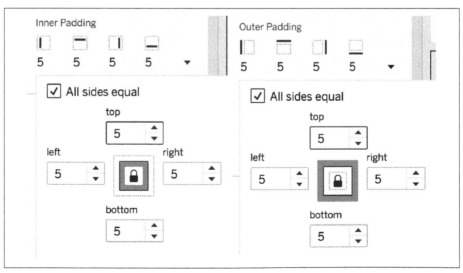

Your final legend should look like this:

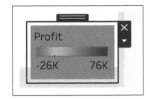

Discussion

Padding is useful for adding whitespace around each individual element. The difference between inner and outer padding is as follows: inner padding adds pixels to the inside of the element (everything inside the border or background), and outer padding adds pixels to the outside of the element (everything outside the borders or backgrounds). Here is a before and after image of no padding, inner padding, and inner and outer padding:

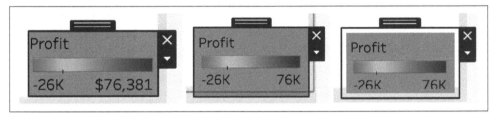

When you are adding objects to containers, Tableau defaults to outer padding of 4 pixels on all sides of the sheets or objects. However, when using floating objects or containers themselves, Tableau has zero padding on the objects.

Item Hierarchy

"Item hierarchy" is at the bottom of the Layout tab in a dashboard. It can be used to organize your dashboards and show the containers or worksheets.

Problem

You want to find the horizontal container to rename it.

Solution

1. In "Item hierarchy," click through until you find the sheets inside the horizontal container:

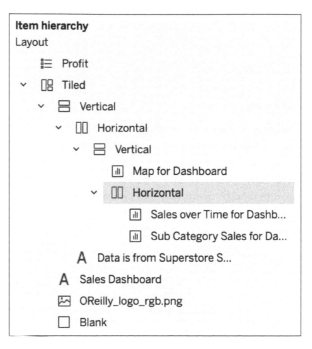

2. Select the container; you can see it is selected in the dashboard by the blue box surrounding the charts:

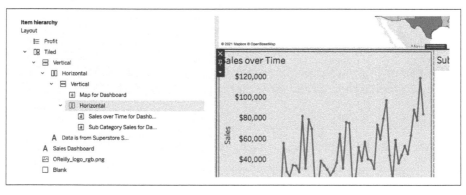

3. Right-click the container in the hierarchy to rename the dashboard item:

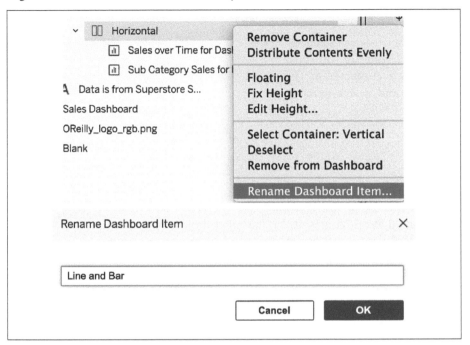

Once you have named your dashboard item, it will appear in the hierarchy with the new name:

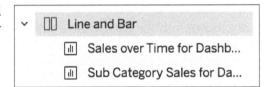

Discussion

The "Item hierarchy" is important in understanding how your dashboards are being built. When you start to drag sheets onto the dashboard, the hierarchy automatically starts with a Tiled container. This is a mix of both horizontal and vertical containers. There could be many nested containers under the tiled one, which can cause confusion when looking back in the hierarchy.

If you are happy with the layout of your dashboard, I recommend removing containers
that are not necessary to elimi-
nate confusion. To do this,
right-click a container in the
hierarchy and select Remove
Container:

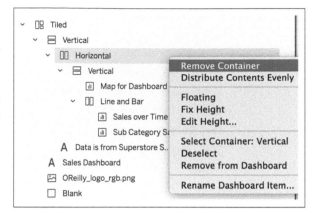

You will also notice on this particular hierarchy that the
color legend Profit comes above the Tiled option. This is
because the element is floating:

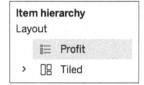

Tiled, Containers, and Floating are three ways you can build dashboards. For more
information, see Recipe 19.3.

Summary

Congratulations. You have successfully built your first dashboard. You should now
understand how to build a simple dashboard with simple interactions. To make your
dashboards more complex, go to Chapter 19.

Part II Conclusion

You are now at the end of Part II. You have built some of the basic charts within
Tableau; what you have learned so far will put you in a good position to get started
with your own data and analysis. Part III digs deeper into some of the other fundamental chart types that you can build within Tableau.

Broadening Your Data Viz Knowledge

CHAPTER 8

Square

Using the Square option on the Marks card, you can create different types of charts, including previously mentioned highlight tables (Recipe 4.3). This chapter will look at several charts covering the Square mark type, which in some instances means rectangles.

8.1 Treemaps

One type of chart using the Square mark type is a treemap. A *treemap* can be used to display part-to-whole or hierarchical relationships using rectangles or squares as the area of the measure.

Problem

You want to show the part-to-whole relationship between categories for sales.

Solution

1. Drag Sales to Size (1) and
   Category to Color (2):

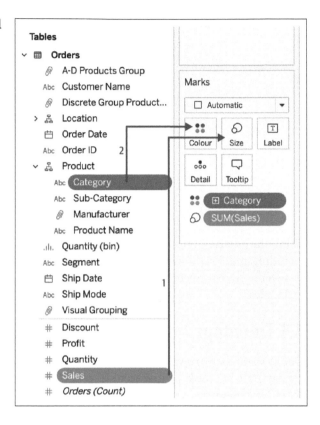

2. To distinguish between the different squares, I advise adding labels. You will need to drag the same two fields from the tables to the Label shelf. The order you drag the fields onto the Marks card depends on whether you want category or sales to appear on the label first. In this example, we put category (1) and then sales (2) onto Label:

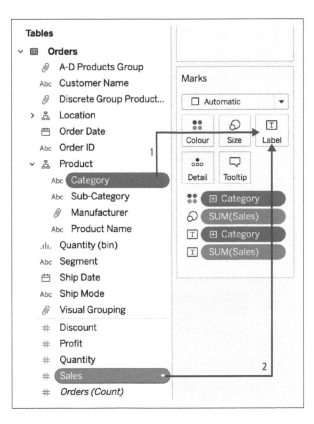

Editing Labels

If the label is still incorrect, you can manually edit it by clicking the Label property. This will bring up a box for you to edit different elements of the label. If you select the three dots next to the Text option, you can specifically edit what appears on the label:

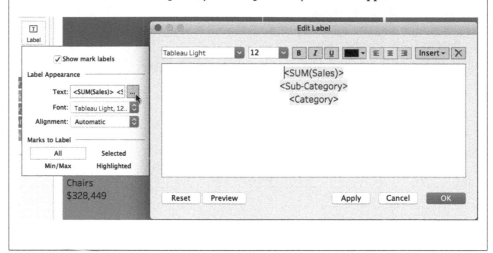

Your final view should look like this:

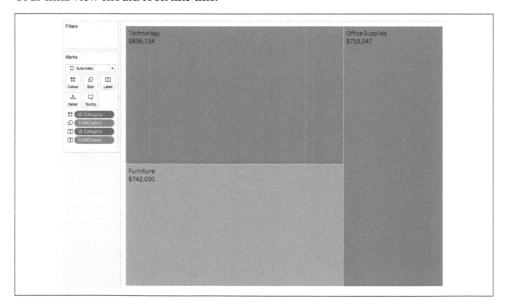

Discussion

Treemaps are good at showing a part-to-whole relation-ship, and they have good visual appeal. However, the user could take longer to interpret this chart type over a bar chart, because the human eye can interpret the pre-attentive attribute of length better than the size attribute of an area. A treemap orders the segments based on highest to lowest, with the largest/highest sections being in the top-right area.

If you want to show how the subcategories fit into these categories, all we need to do is click the plus sign on Category that is on the label, which will drill down to Sub-Category:

 If you don't have the Category hierarchy, you will need to drag Sub-Category onto the Label shelf.

Now the view looks like this:

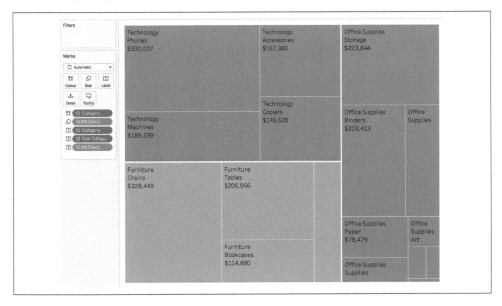

Once you have drilled down to Sub-Categories, you can also add Sub-Category to Color, which will then color your areas by category with different shades per subcategory. To do this, duplicate Sub-Category to the Detail property. You can do this by either dragging Sub-Category from the Data pane or pressing Ctrl (Command on Mac) and dragging the Sub-Category field onto the Detail property:

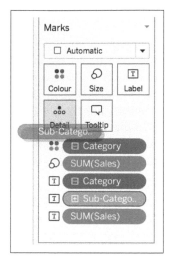

You can now click the three dots by the Sub-Category field that is on the Detail shelf and change the shelf to Color:

Your view will now look like this:

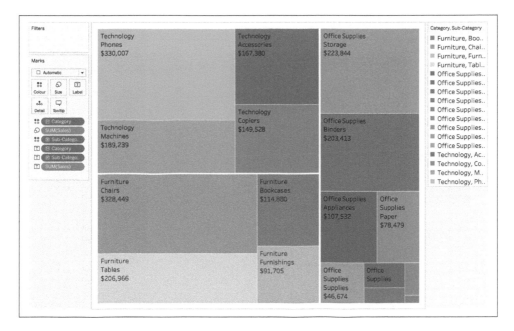

Having two dimensions on the Color property allows you to segment any chart. However, this does make the color legend unreadable, and you could end up with a lot of colors, depending on the fields you use.

Because of the order of the Marks card, this still groups each area by the total amount of sales per category. However, if you want to see the highest amount of sales regardless of category, you can switch the order on the Marks card so that Sub-Category is above the first Category on the Marks card:

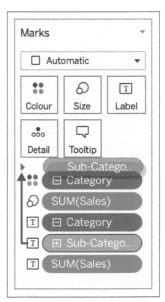

Your view now looks like this:

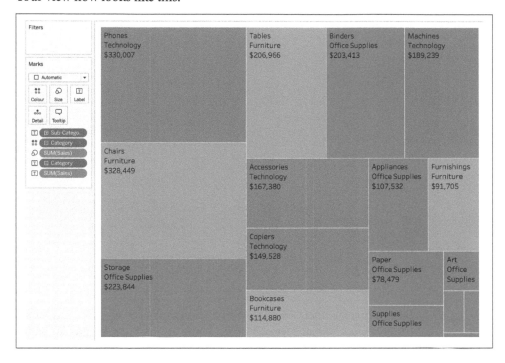

An advantage of using treemaps is you can use multiple measures. For example, you can use sales as the size, but you can color by a different measure, like profit. If we revert back to where the treemap is grouped by category and add Profit to Color instead of Category, your view will look like this:

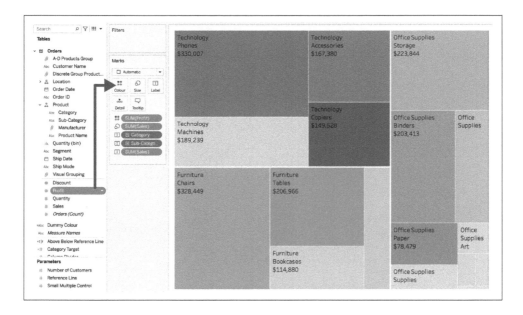

I don't recommend sizing by a measure that has negative numbers, because to show the subcategory, you would have to use absolute values, which could be misleading.

8.2 Calendar Heatmap

In Recipe 4.3, I showed you how to create a text highlight table. This calendar heatmap follows similar principles, and you will be using dates to give a calendar feel to your heatmap.

Problem

You want to display a single month of a calendar to show the number of orders shipped by day.

Solution

1. To start with, we want to filter to a single month. You can do this by dragging Ship Date to Filters, selecting Month/Year, and choosing your month.

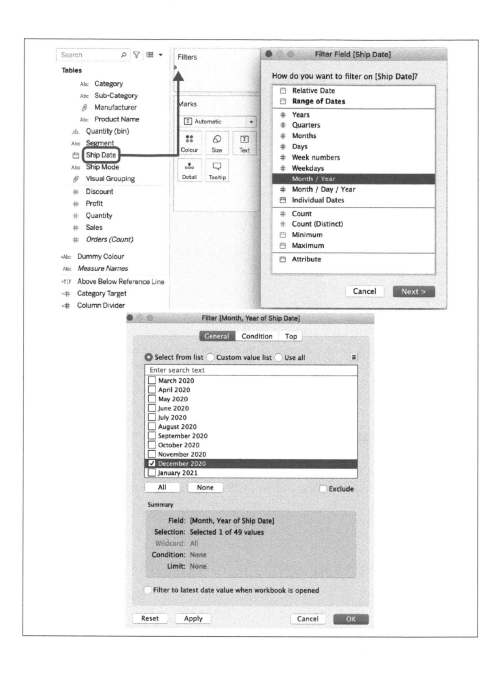

2. Now we need to build our calendar. Start by right-clicking (Option on Mac) and dragging Ship Date to Rows:

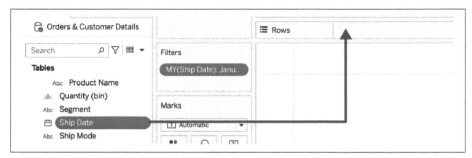

3. By right-clicking and dragging (pressing Option and dragging for Mac), you should automatically get this Drop Field date pop-up box. Select WEEK(Ship Date):

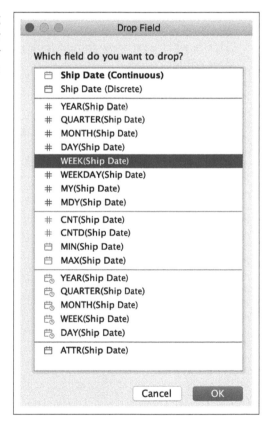

4. Next, we need our weekdays, so right-click and drag another ship date, but this time to Columns. When the pop-up box appears, select WEEKDAY(Ship Date):

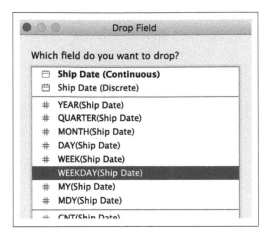

This is what your view should look like, week numbers going down and weekdays going across:

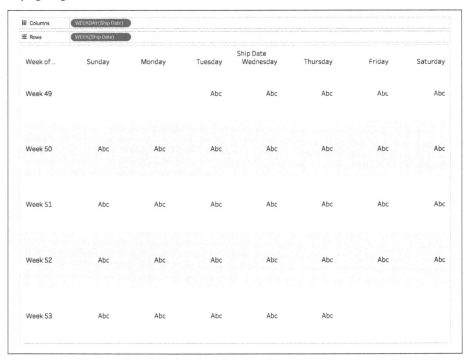

5. The next step is to add the actual day number to the view. Right-click and drag another ship date to Text on the Marks card and select DAY(Ship Date):

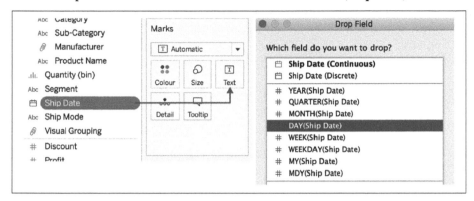

6. To move the day number to the top right of the area, go into Label and change the alignment to horizontally right and vertically top:

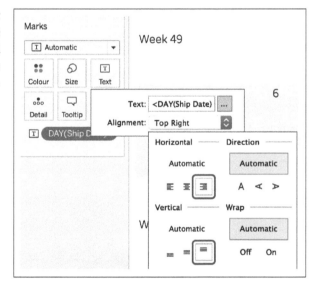

7. Finally, we want to see the number of orders shipped on those dates. To do this, right-click and drag Order ID to Color on the Marks card and select CNTD. This will give you the Distinct Count of Order IDs:

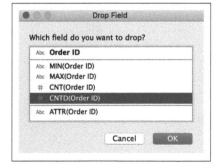

8. Due to the nature of what we have built, we need to change the mark type back to Square; because the Mark Type was on Automatic, it defaulted to the Text mark type. Click the drop-down and select Square:

Your final view should look like this:

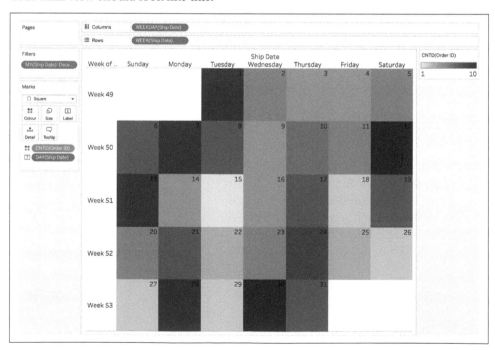

Discussion

By creating this calendar in Tableau, you can quickly see how many orders are due to be shipped on a certain date, meaning you can plan resources as required. A calendar heatmap shows you hot and cold spots in a monthly view.

Since we added the Month/Year filter, we can give the user the ability to change which date they are looking at. Right-click the MY(Ship Date) filter and select Show Filter; this will add the filter to the view:

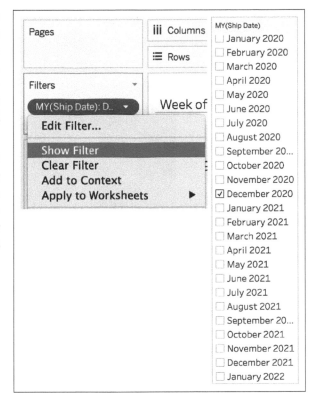

8.3 Marginal Bar Charts

A *marginal bar chart* is a mix of heatmaps and bar charts stitched together on a dashboard.

Problem

You want to see which weekday and week was most popular for shipping orders.

Solution

1. Duplicate or re-create Recipe 8.2 so you have one view that looks like this, and the sheet is named appropriately to find it when creating a dashboard:

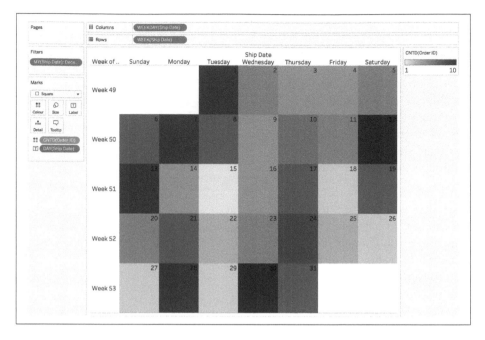

2. The next step is to create a vertical bar chart to show the number of orders per weekday, regardless of week.

On a new sheet, right-click and drag Ship Date to Columns and select WEEK-DAY. Then right-click and drag Order ID to Rows and select CNTD:

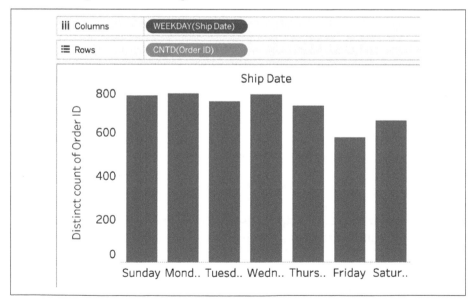

3. Also add CNTD(Order ID) to Color:

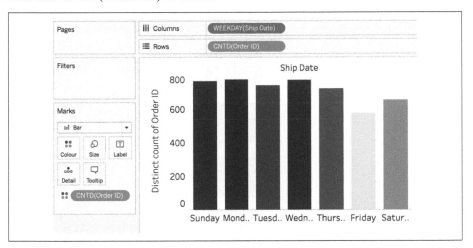

4. For the second bar chart, we can duplicate the WeekDay Histogram bar chart sheet. But this time, we want to create a horizontal bar chart that looks at weeks irrespective of weekdays:

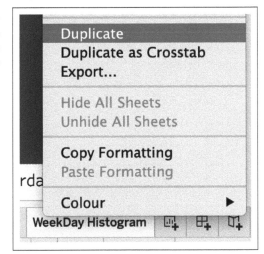

5. On this newly duplicated sheet, use the Switch Axis button to help us switch to a horizontal bar chart instead of a vertical one:

6. Then we need to change the weekday field on Rows to WEEK instead. You can right-click this field and select Week Number:

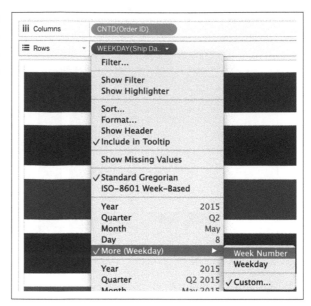

This sheet should now look like this:

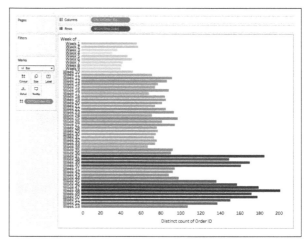

7. We need to apply the Month/Year filter from Calendar HeatMap to the WeekDay and Week Histograms. To do this, go back to the calendar heatmap sheet, right-click the filter, and select Apply to Worksheets > Selected Worksheets:

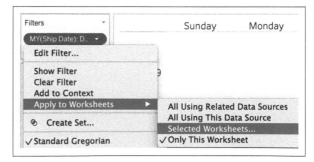

8. This will give you a pop-up box; select the specific worksheets you want to add the filter to:

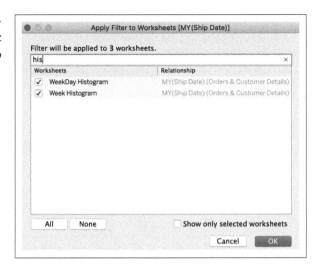

Adding this filter to the histogram sheets will now make them look like this:

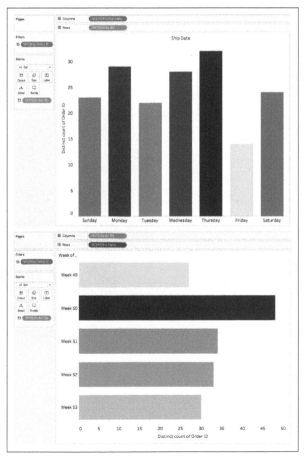

9. Now we need to stitch them together on a dashboard. Create a new dashboard and start by dragging in Calendar HeatMap:

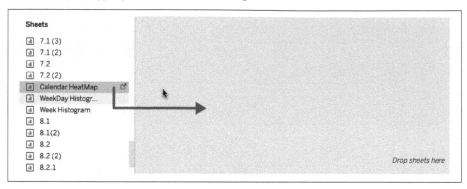

10. Add the weekday bar chart above our calendar heatmap:

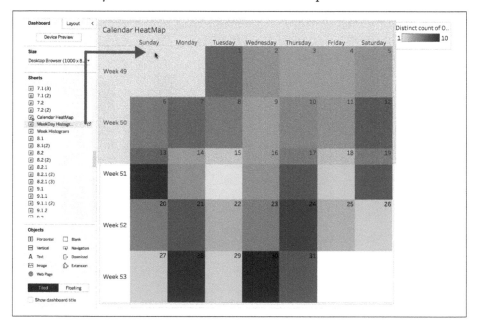

11. Add our week number horizontal bar chart to the right-hand side of the calendar heatmap:

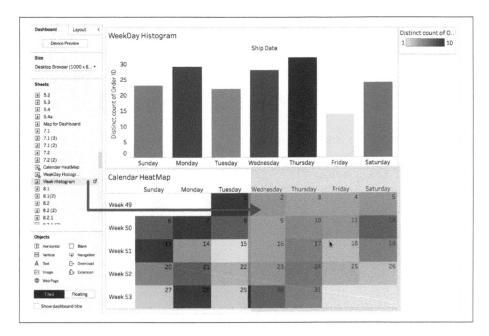

12. Now that we have all of our sheets on the dashboard, we need to clean up the dashboard and make sure everything aligns. We can add a blank object to the right-hand side of the weekday bar chart, which will allow us to align the weekdays:

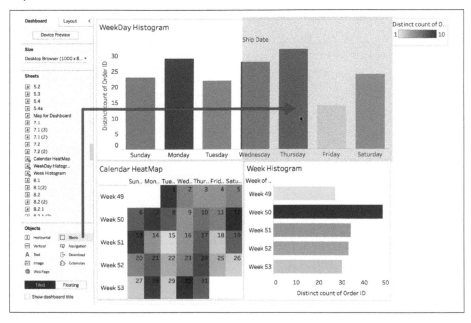

13. From here, hide the titles of each worksheet by right-clicking the title of the worksheet:

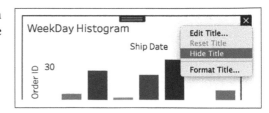

14. Unshow the axis and headers of the bar charts, by right-clicking either the axis or header:

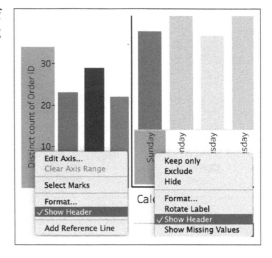

15. Repeat steps 13 and 14 for the Week Histogram sheet.

16. Hide column and row field titles:

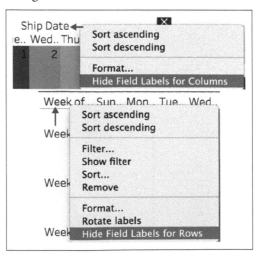

Now your view should look like this:

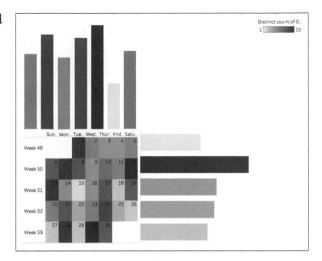

17. As the calendar heatmap is the central chart, we should make it bigger. You can do this by hovering between the charts and then dragging up/down or left/right:

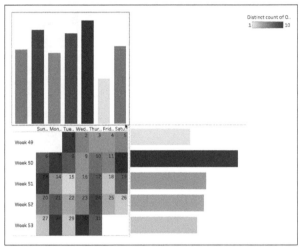

Once you are happy with the size of the calendar, your view should look like this:

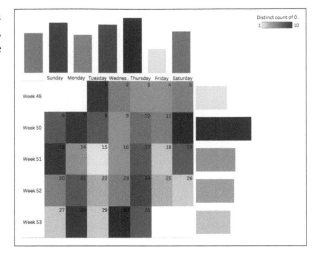

18. Notice that the weekday histogram doesn't align properly with the heat-map. Add a blank and adjust the size. The view should now look like this:

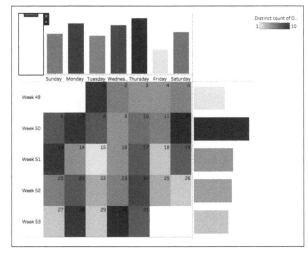

This recipe uses blank objects for whitespace, but you can also use inner or outer padding.

19. We should add labels to our bar charts to provide context for the scale of values. You can do this from the dashboard by selecting the sheet and clicking the T icon on the toolbar:

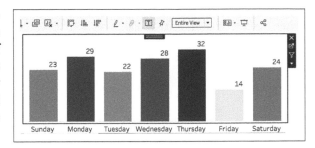

20. Now that we have context, you can remove the color legend, and your final view should look like this:

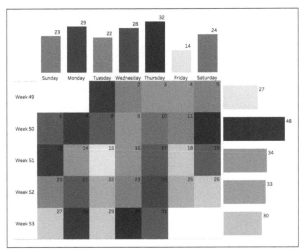

Discussion

Marginal bar charts are used to show the higher level of aggregation from the heat map. As mentioned in the recipe, each bar chart shows the level of detail for either weekday without week numbers or week numbers without the weekday. Arranging these charts into marginal bar charts on dashboards gives each week number and weekday a grand total as a bar chart instead of text numbers.

Marginal bar charts are useful as you can see weekday trends on the top bar chart as well as individual busy days in the calendar heatmap.

Summary

The Square mark type is useful for building treemaps and heat maps in Tableau. You now know how to build both, as well as how to combine them onto a dashboard to make a marginal bar chart. Heatmaps are especially useful when trying to convert your users from tables in Excel to something more visually appealing.

CHAPTER 9

Tooltips

Tooltips are the information that appears when you hover over a mark on a visualization; they are automatically available on each chart you build. Primarily, they help provide the information in the chart in a text format. Tooltips can also be used to provide additional information that isn't included in the chart.

This is what the default tooltip looks like:

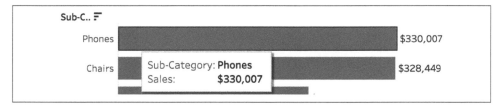

9.1 Basic Tooltips

Basic tooltips can be left unchanged and not formatted, but sometimes you can end up with extra fields that you might not need in a tooltip. You can edit a basic tooltip in a variety of ways, from stating the facts to forming sentences—this depends on the use case.

Problem

You want to edit and format the default tooltip as well as add information.

Solution

1. Start with a horizontal bar chart showing Sub-Category by Sales, sorted in descending order (see Recipe 3.1).

2. On the Marks card, click the Tooltip property:

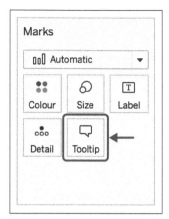

An automatic pop-up box pens that will look like this:

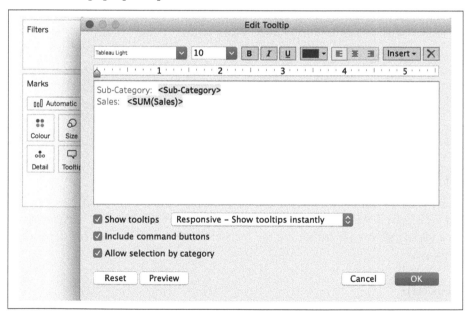

 The gray shaded text is dynamic per individual mark you hover over, whereas the unshaded text is static and will appear the same on every mark.

3. At the top of this box, you have options to edit the font, size, and some format of the text for the tooltip (1). You can also change the color and alignment of the text (2):

4. The Insert drop-down allows you to insert a variety of options. The majority of the options are always available and are the default in Tableau, i.e., Data Source Name or Page Count. The seventh section is where your fields appear that are in your active sheet.

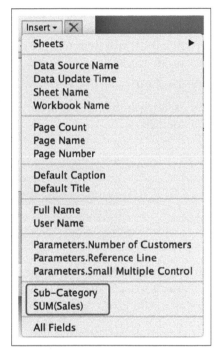

5. Now let's change the text to **\<Sub-Category\> sales were \<SUM(Sales)\>**:

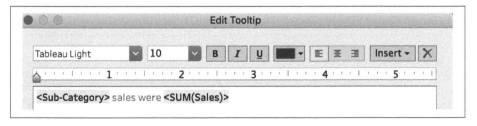

6. Preview how this will look by using the Preview button:

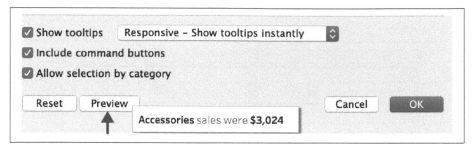

7. Finally, if you want to reset the tooltip back to the default, click the Reset button inside the tooltip.

8. Click OK.

9. When you hover over a bar, you should now see the sentence we created in the tooltip:

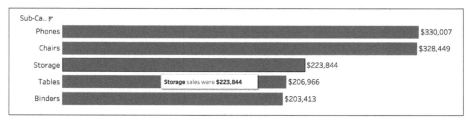

10. We also want to see what the profit was for each of these subcategories. Add Profit to Color:

11. If you hover over the bar chart, you will notice that profit doesn't automatically appear on the tooltip. That is because we have already formatted it:

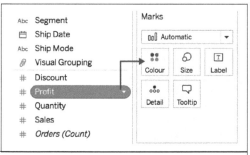

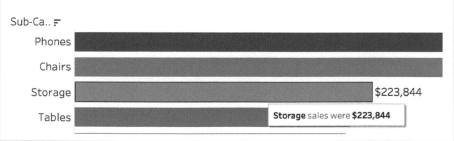

12. To add this into the tooltip, we have to edit the tooltip (1), click Insert (2), and select SUM(Profit) (3). Then click OK:

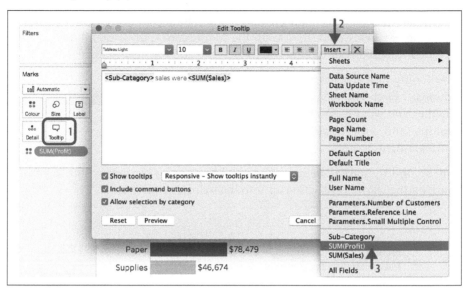

13. Finally, add an extra sentence to the tooltip:

Now when you hover the mouse over the bar chart you will see both Sales and Profit in tooltip:

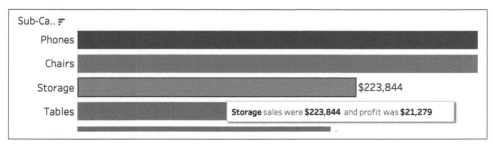

Discussion

Tooltips are great to clarify information when you hover over the bar chart. You can add as many dimensions and measures to the tooltip as you'd like by dragging and dropping the fields onto Tooltip on the Marks card, but as mentioned, if you have already formatted your tooltip, you will have to add the field manually.

The Tooltip property has other options. First is choosing whether you want to show the tooltips, and how responsive you want them to be:

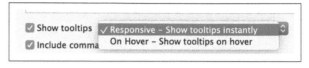

The second option is "Include command buttons." This means that if you select a mark on the chart, it will come up with the following options:

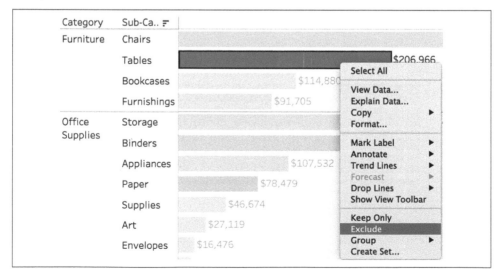

These *command buttons* allow your user to filter out data, group data, and several other things. I recommend turning off this option to avoid accidental filtering.

If you do turn off the command options, a user can still right-click a mark and have the option to exclude the mark from the view:

Turning off "Include command buttons" is also recommended for security reasons. When publishing to Tableau Server or Tableau Online, if the permissions are not set correctly, the command buttons allow users to access the underlying data.

The final option is "Allow selection by category" when multiple options are available in your view. For example, if we add Category to the view and then add it to the tooltip, it will look like this:

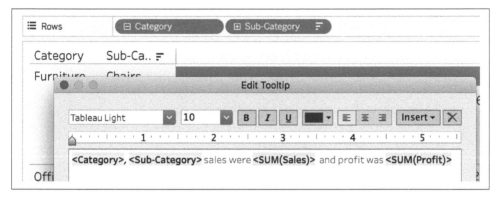

When you now select a mark inside the Furniture category and hover over the word "Furniture," it becomes underlined:

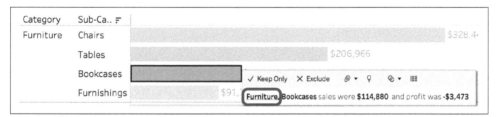

Then when you click Furniture, all the other furniture subcategories will be highlighted, and with the command buttons switched on, there could be for more accidental filtering:

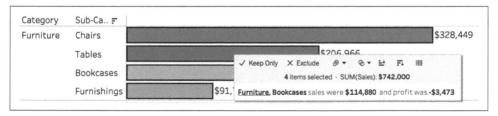

A basic tooltip allows you to show textual information relevant to your chart. However, what if you want to add something more visually appealing inside the tooltip?

9.2 Viz in Tooltips

Viz in Tooltips allows you to add another visualization to the tooltip to add further information and/or context.

Problem

You want to have a quick look at subcategory sales when you hover over a category.

Solution

1. Create a new sheet with a horizontal bar chart using Category and Sum of Sales.

2. We need to have another sheet ready to go into the tooltip. We can use the chart created in Recipe 9.1.

3. On the first sheet (Category by Sales), edit the tooltip:

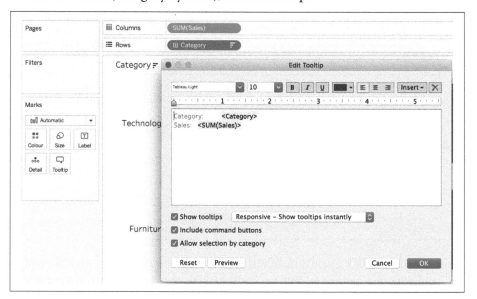

4. Now we need to insert a new sheet into this tooltip. Click the Insert button, and choose Sheets:

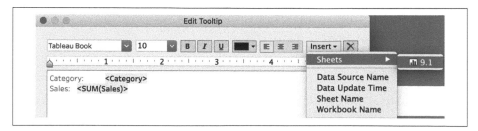

This will insert a markup underneath the original tooltip like this:

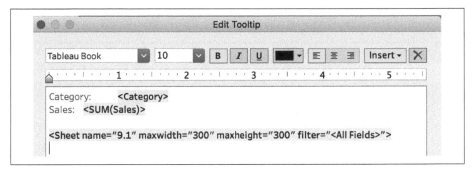

5. Once you click OK and hover over a category, it should look like this:

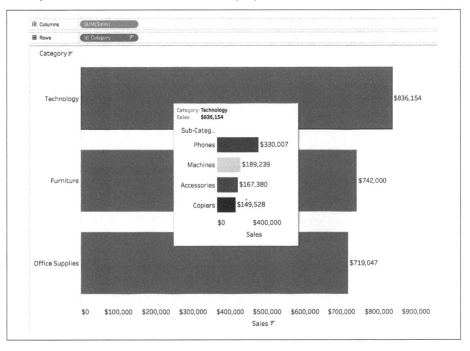

Discussion

Viz in Tooltips are very useful for showing extra information on a visualization. A downside is that the Viz in Tooltip doesn't have its own tooltip to find out what other numbers are in that visualization.

In the markup for the Viz in Tooltip, you can change various elements to enhance the visualization being shown. For example, you can change the size of the Viz in Tooltip by changing the `maxheight` and `maxwidth` options to different pixels:

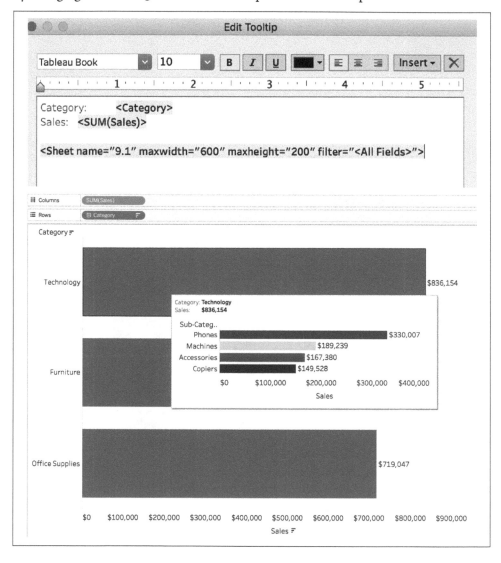

Another element in the Viz in Tooltip markup that you can change is `<All Fields>`. This is the field that is filtering the visualization. For example, if you had an extra level of detail (segment) in your view. All fields would continue to filter the subcategory Viz in Tooltip view to both the category and the segment you have hovered over. If you change the `<All Fields>` option to `<Segment>`, this will then filter to the only segment level of detail, like the following:

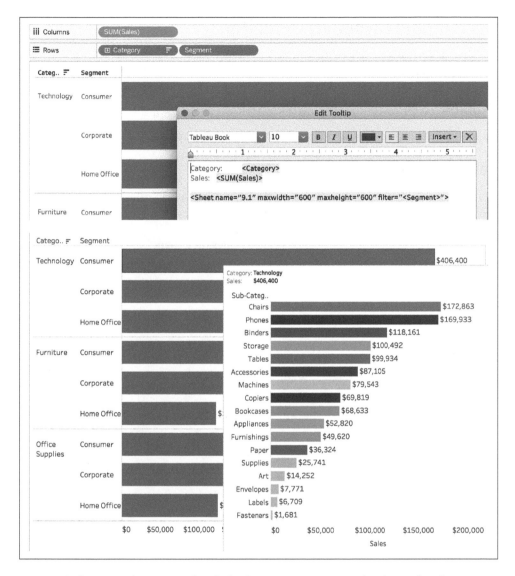

Viz in Tooltips can be accessed only by hovering over a mark. If you decide to print the visualization, information within the tooltip will not be printed. So, if the information inside the Viz in Tooltip is mission critical, you should think about including it on a dashboard rather than in the tooltip.

Having a single Viz in Tooltip is powerful, but what if you could have a mini dashboard inside the tooltip?

9.3 Multiple Layout Viz in Tooltip

A multiple layout Viz in Tooltip allows you to use multiple sheets stacked or side-by-side inside the tooltip.

Problem

You want to see sales for each subcategory and state, over time, for each category.

Solution

Duplicating Recipe 9.2 means we already have subcategory sales inside the tooltip. Now we need to add state-level sales and sales over time side-by-side in the tooltip.

1. Create a new sheet and show the states by sales on a map:

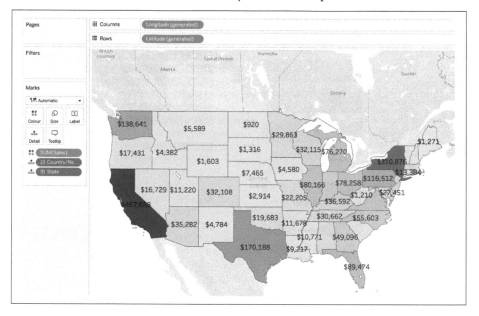

2. Create another new sheet with sales over time:

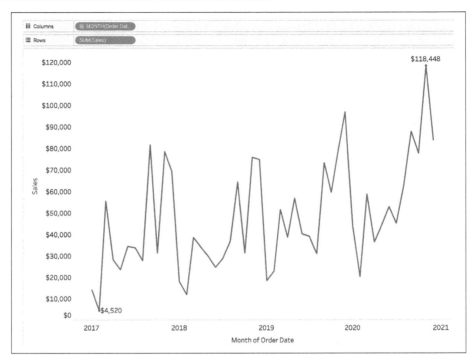

3. Go back to our original sales by category sheet and edit the tooltip:

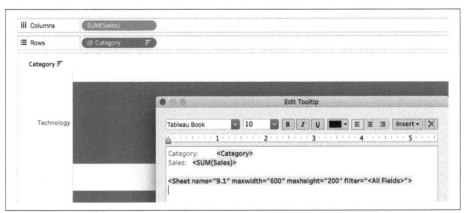

4. Following previous steps, we want to add a new sheet inside this tooltip. First, put your cursor at the end of the first markup for the Viz in Tooltip. Second, click the Insert button and find the map of sales by state:

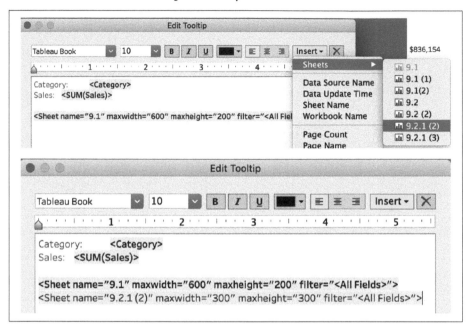

5. Now to get the map of sales by state and the sales over time sheet next to each other, you need to place your cursor directly next to the second Viz in Tooltip markup:

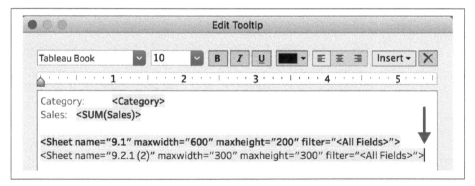

6. Go to Sheets and find the sales over time sheet you just created:

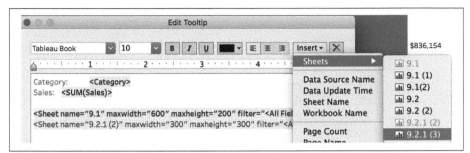

Your final markup should look like this:

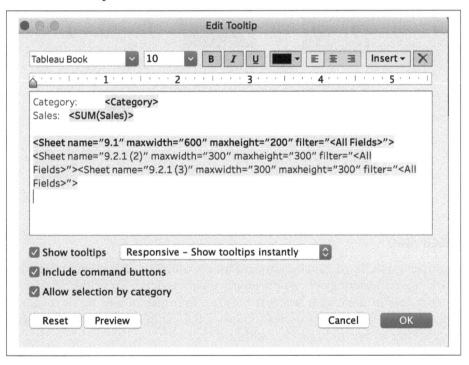

7. Now when you click OK and hover over a mark, your Viz in Tooltip should look like this:

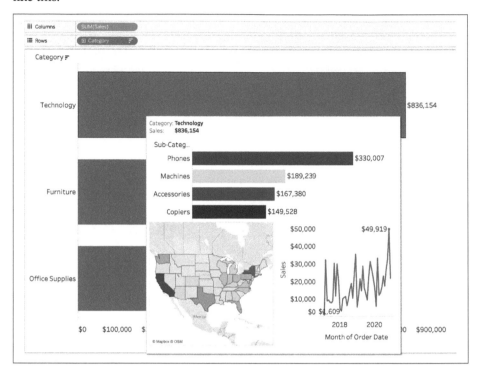

Discussion

Multiple layout Viz in Tooltips are good for providing a quick, deeper insight into the mark you are hovering over. As previously mentioned, the Viz in Tooltip doesn't have its own tooltip then, so it will just be static information and not accessible when printed. Therefore, if this information inside the Viz in Tooltip is very important, it should have its own sheet on the dashboard.

Summary

Editing tooltips is something that should be done at the very end of your worksheets or dashboard creation. Otherwise, you will have to go back and edit them again if you change something. Make sure you edit your tooltips, as they can make a big difference to your end user.

CHAPTER 10

Area Charts

An *area chart* is a line chart with the area beneath the line shaded. Tableau has a variety of options for area charts that are dependent on the dimensions and measures you use.

10.1 Basic Area Charts

A basic area chart uses ordinal data and a measure, for example, continuous dates and sales. This is very similar to a continuous line chart, but with the area shaded.

Problem

You want to see an area chart for sales by continuous months.

Solution

1. On a new worksheet, double-click Sales, which automatically adds the Sum of Sales field to Rows.

2. Right-click and drag (Option-drag on Mac) Order Date to Columns and select continuous months:

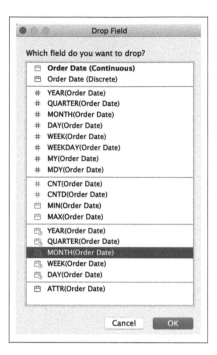

3. The final step is to change the mark type from Automatic to Area. Click the drop-down menu on the Marks card (1) and select Area (2):

Your final view should look like this:

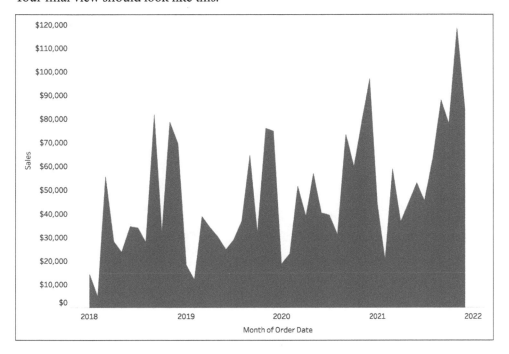

Discussion

A continuous area chart is exactly like a line chart but with the shaded area beneath the line. When creating this type of chart, I like to add a border on the area chart to make the marks easier to see.

To add a border, click Color on the Marks card and select Border:

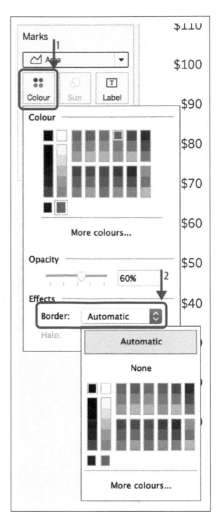

With an added black border, this is what your view would look like:

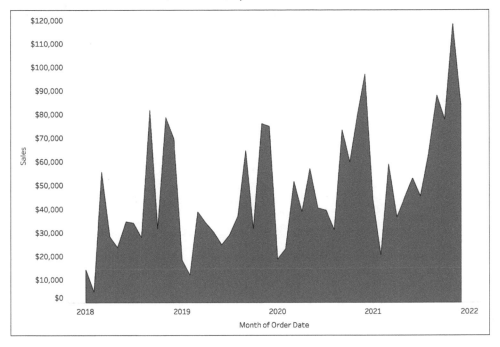

This is how to create a continuous area chart, but what if you want to split the view by a dimension?

10.2 Stacked Area Charts

A *stacked area chart* allows you to visualize the dimensional proportions by stacking the volumes on top of each other, but with the added benefit of being able to see the total value of the dimensions.

Problem

You want to see the volume of each segment over time.

Solution

1. Duplicate the sheet in Recipe 10.1.

2. Double-click Segment, which will automatically add it onto Color on the Marks card:

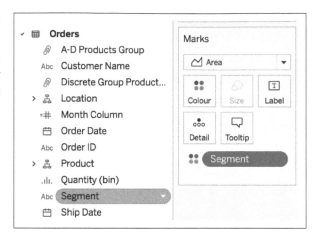

Your final view should look like this:

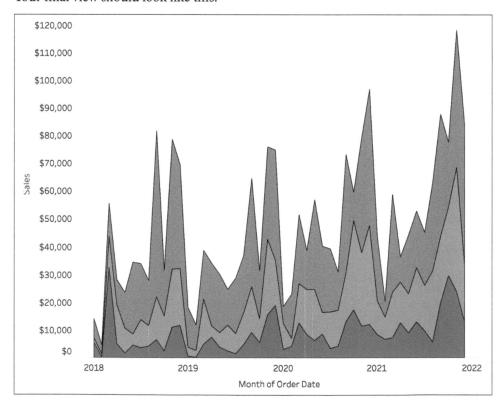

Discussion

With a stacked area chart, you can see the volume of each dimension and how that changes over time, but it can be difficult to compare each of the months. You can also see the total value by looking at the top of the areas covered, as well as what contributes to that total of each month.

 When hovering over a particular mark, Tableau will give you only the value for the segment, not the total value.

If you want to be able to hover over the total sales regardless of the dimension split, you can create a dual axis with a line chart that will allow for this tooltip at the top of the chart. To create this, start by adding another Sum of Sales field to Rows (1), and on the second Marks card (2) remove Segment and change to Line (3):

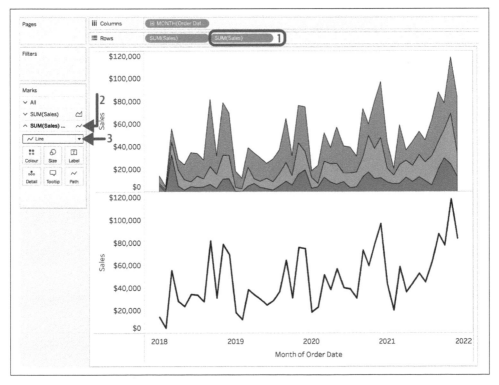

Finally, right-click the second Sum of Sales field and select Dual Axis. Then, don't forget to synchronize your axis, by right-clicking the right-hand axis and selecting Synchronize Axis. Now, right-click the right axis and deselect Show Header:

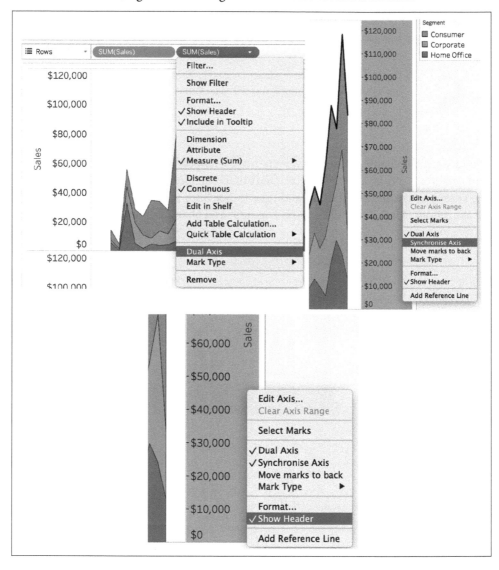

Doing this will allow you to hover over the black line to see the total amount of sales, rather than split by segment, and your final view should look like this:

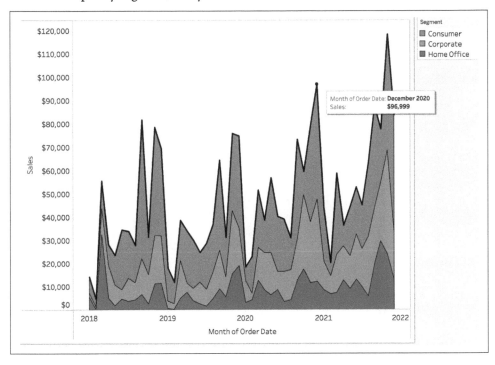

The difficulty with this view is that you cannot compare directly to anything other than the bottom dimension. Because the colors don't have a common baseline, you can't really see how the middle or top segment is performing. This is where a Percent of Total area chart will help.

The segments are stacked by default in Tableau, but you can turn off Stack Marks, which was mentioned in Chapter 3. To do this, choose Analysis > Stack Marks > Off:

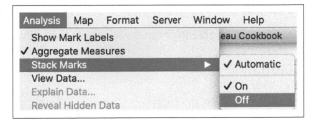

This will make the areas stack behind each other instead of on top:

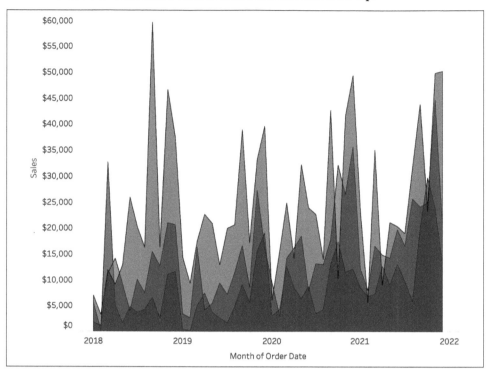

This, however, is not the best practice because of the overlapping colors and marks.

10.3 Percent of Total Area Charts

Problem

You want to be able to compare sales values as a Percent of Total area chart.

Solution

1. Duplicate Recipe 10.2.
2. Right-click the SUM(Sales) field in Rows and choose Quick Table Calculation > Percent of Total:

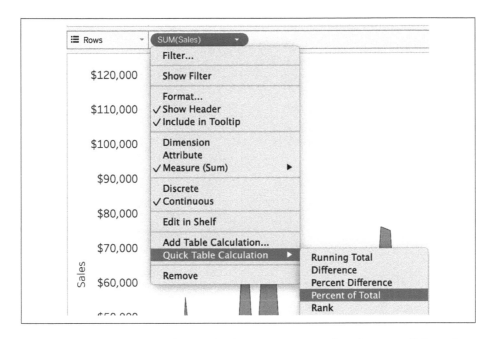

The default computation of this calculation is percent of total across all the values in the view:

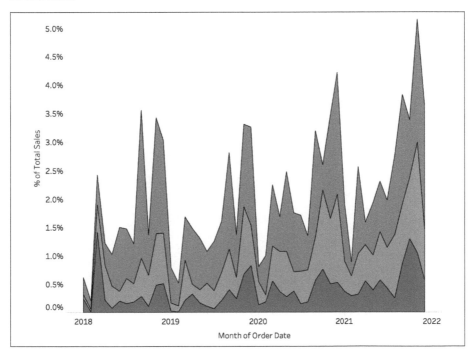

3. We want it to compute using only the segment, so right-click the SUM(Sales) field in Rows and choose Compute Using > Segment:

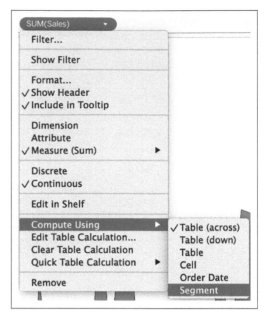

Your final view should look like this:

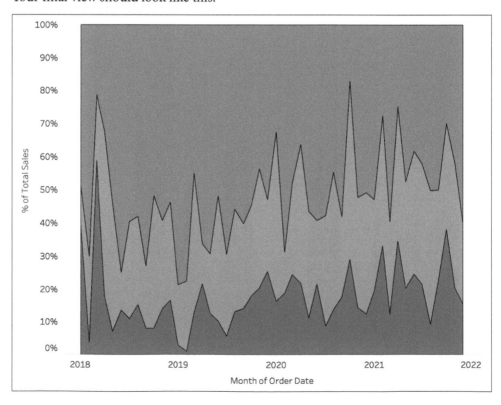

Discussion

Using a stacked Percent of Total chart illustrates a part-to-whole distribution. Using this type of chart can still pose some challenges when reading, but it does show how the percents of total sales are changing over time. The alternative to using an area chart is to change the mark type to Line; this can make it easier to compare values.

10.4 Discrete Area Charts

The differences between continuous and discrete area charts are the same as the differences between a continuous and discrete line chart (Chapter 5). Using discrete dates instead allows for segmentation of the individual date parts.

Problem

You want to segment the dates to show a stacked area chart.

Solution

1. Press Ctrl (Command on Mac) and select Order Date and Sales from the left side:

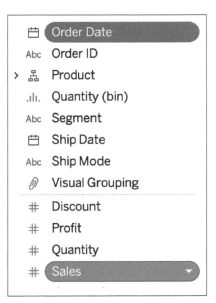

2. Using the Show Me menu, select the discrete area chart:

Your initial view should look like this:

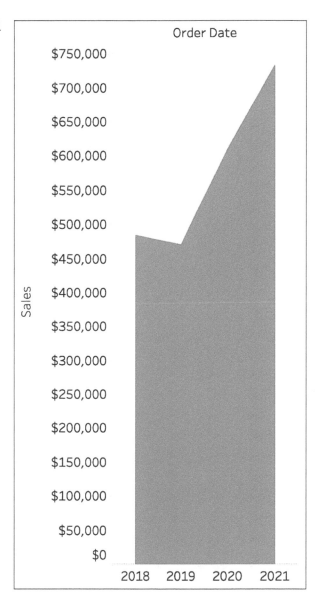

3. If you want to drill down to month, right-click and drag Order Date to Columns and select discrete months:

Your view should now look like this:

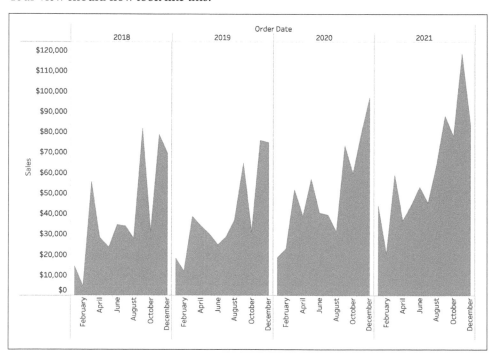

Discussion

Using discrete area charts allows the segmentation of the individual dates, which allows you to spot seasonal trends in the data.

10.5 Pareto Charts

A *Pareto chart* is a different type of area chart. It is named after Vilfredo Pareto, who made the observation that 80% of the land was owned by 20% of the population. In general terms, it could mean 80% of consequences should be made up of 20% of the causes. This can be translated to 80% of sales should be made up of 20% of the products. This can be translated into many scenarios.

Problem

You want to see what percentage of products account for at least 80% of sales.

Solution

1. Double-click Sales, which automatically adds it to Rows.

2. Double-click Product Name, which automatically adds it to Columns.

3. Make sure the view type says Entire View, and it should look like this:

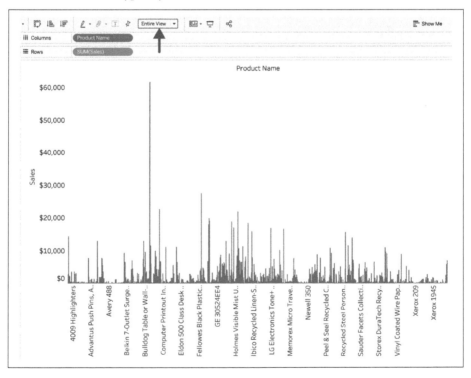

4. Next, we need to sort the products by the Sum of Sales. You can do this by clicking the Sort button in the top toolbar:

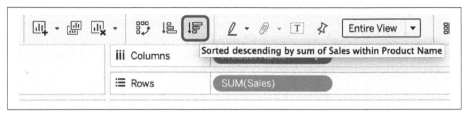

Now your view should look like this:

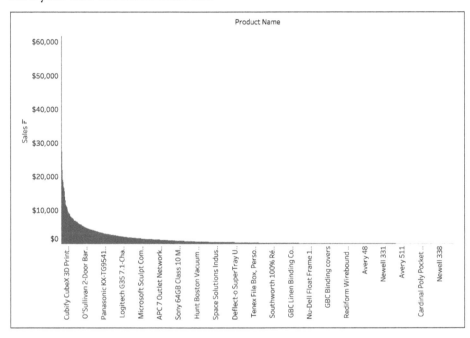

 The next steps need to be followed carefully.

5. On the Sum of Sales field in Rows, we need to add two Table Calculations. Right-click Sum of Sales and select Add Table Calculation:

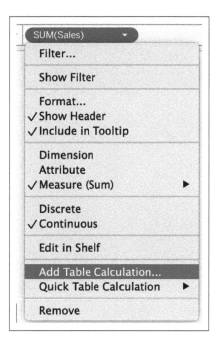

6. This will bring up a pop-up box. Here we first want to select Running Total:

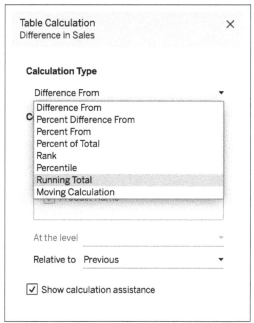

7. Make sure the Specific Dimensions option is selected:

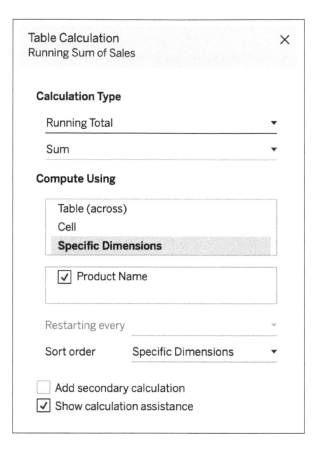

8. The next step is to "Add secondary calculation" (1), which needs to be the Percent of Total (2) with Specific Dimensions selected (3):

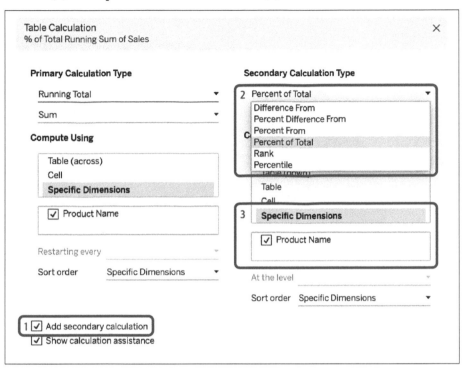

Your view should look like this and now gives us the percent of total sales each product makes up:

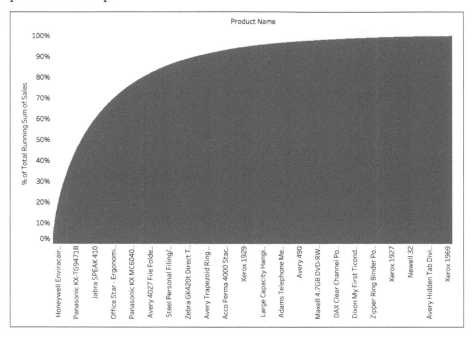

The next part is to get the Percent of Total Products too, which requires repeating steps 6 to 8 but using a different field.

9. First, if we remove the Product Name from the view, we will get red error fields on the Sum of Sales. This is because we have specifically set Compute Using to Product Name. So, we need to duplicate this field onto Detail on the Marks card:

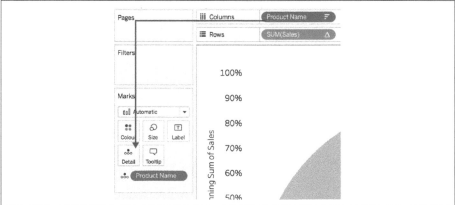

10. Notice that this field doesn't have the sort icon next to it. The sort option is essential for building the Pareto. To add the sort to this field, right-click, select Sort, and choose by Field, SUM(Sales):

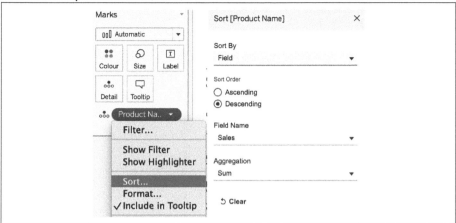

11. To add the next set of Table Calculations, we need a measure for the products.

Right-click on the Product Name, and select Measure > Count (Distinct):

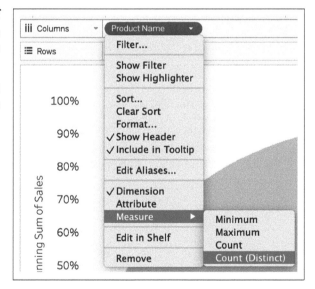

This will look like it has broken your view, but it's because we need to add the Table Calculations:

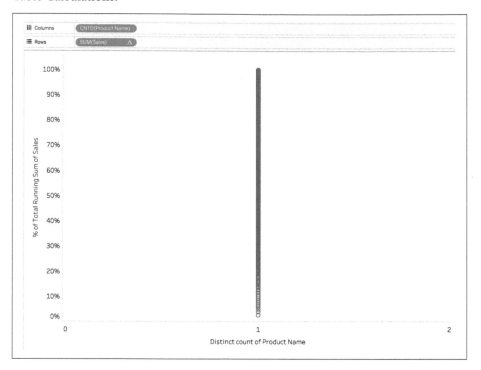

12. Now it's time to repeat steps 6–9. Right-click the CNTD(Product Name) field and select Add Table Calculation:

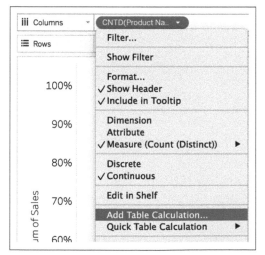

13. Change the type to Running Total and select Specific Dimensions:

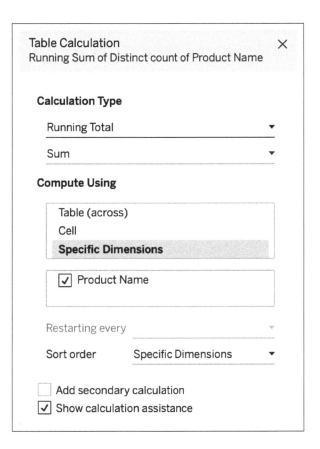

14. Now to add the secondary calculation of Percent of Total by Specific Dimensions:

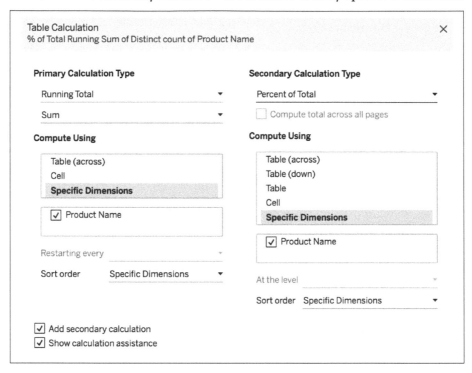

15. The last step for the Pareto is to change the mark type to Area:

Your view should now look like this:

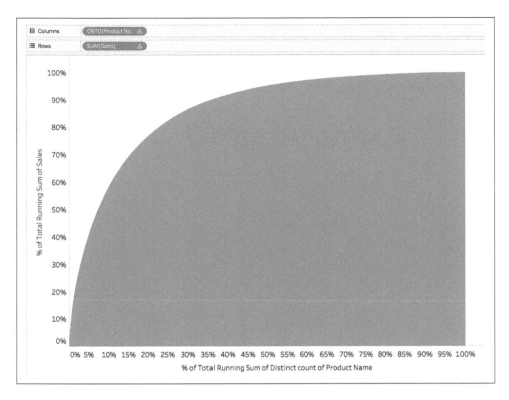

To aid the analysis, I recommend adding constant lines, which allows you to definitively see whether the data falls in with the 80/20 rule.

16. Go to the Analytics pane and drag and drop Constant Line onto the view:

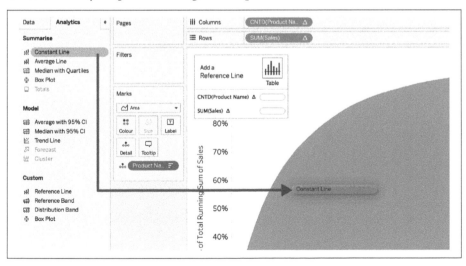

17. This will allow you to add 0.2 to the Product constant line (Rows) and 0.8 to the Sales constant line (Columns):

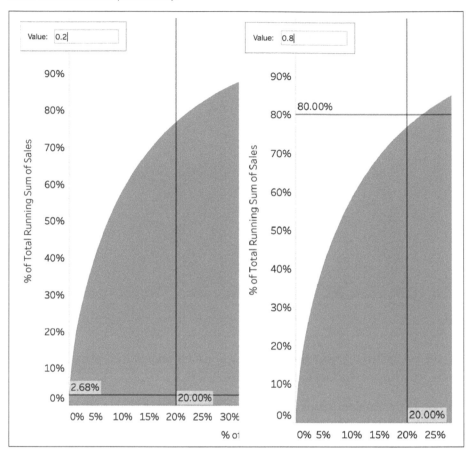

Your final view should now look like this:

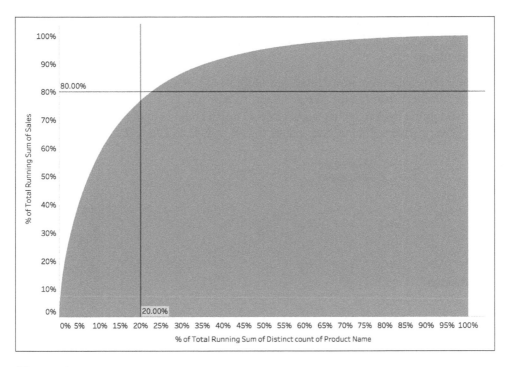

Discussion

In this particular example, 20% of the products account for 77% of the total sales:

For 80% of sales, 23% is made up of the products:

Using this chart type allows you to see how closely your data follows the Pareto rule and to quickly highlight areas of opportunities to improve sales or product development.

Often, Pareto charts are just the percent of total for Sales on the y-axis and a dual axis with a bar chart showing the raw value for each product. To re-create this view, I recommend going back to the view after step 8, which is the running total and percent of total sales by product name, and change the mark type to a Line to create a line chart:

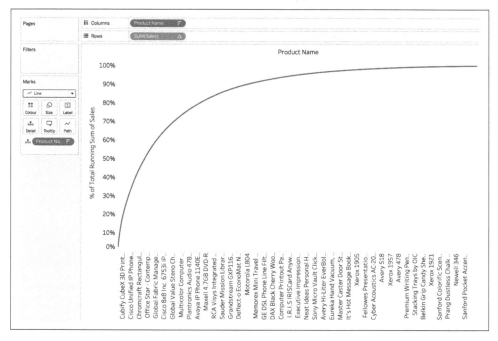

Now we need to add a bar chart showing the individual sales for each product. To do this, we need to create a dual axis with Sales. I have mentioned how you can create a dual axis by adding the secondary field to Rows or Columns and then right-clicking the secondary axis and selecting Dual Axis. You can also drag and drop to the opposite side of the primary axis to create the dual axis:

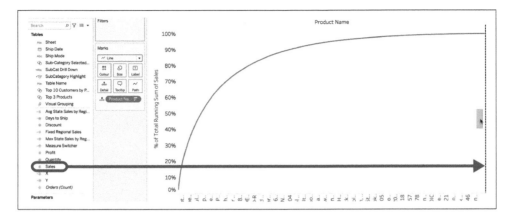

This action creates the dual axis automatically:

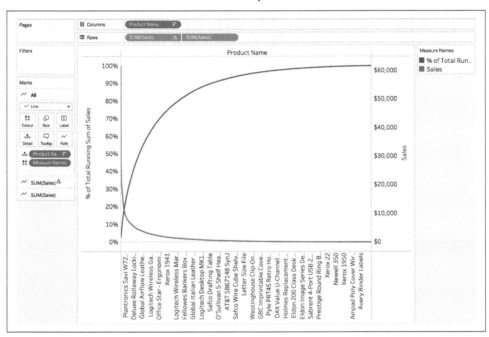

The final step is to change the secondary Marks card to Bar.

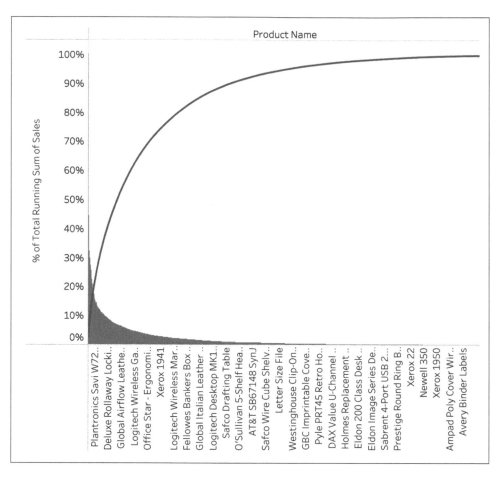

Summary

In this chapter, I introduced the concept of using the Area mark type. Area charts can be used to show cumulative values or percent of totals. Pareto charts are very useful in seeing how closely your data follows the Pareto principle and quickly highlighting areas of opportunities to improve sales or product development.

CHAPTER 11

Circles, Shapes, and Pies

Circle, Shape, and Pie are all mark types on the Tableau Marks card. The chart type you choose will depend on your use case. Throughout this chapter, I'll show you how to use the Circle and Shape mark types and how to create pie charts.

11.1 Scatter Plot

When you want to compare two continuous measures, you might want to use a *scatter plot*. This allows you to see if there is a relationship between the two variables.

Problem

You want to compare sales and profit by customer.

Solution

1. Drag Sales onto Rows and Profit onto Columns:

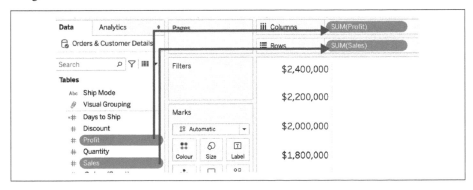

This will show a single mark in the view because Tableau automatically aggregates the view to Sum of Sales and Sum of Profit. It should look like this:

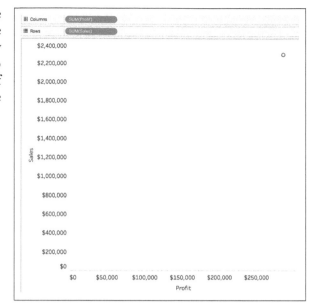

2. To break this down by customer, add Customer Name onto the Detail property of the Marks card:

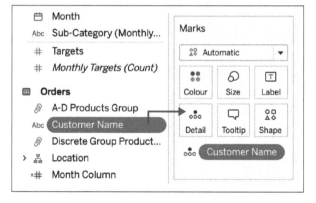

Your final view should look like this:

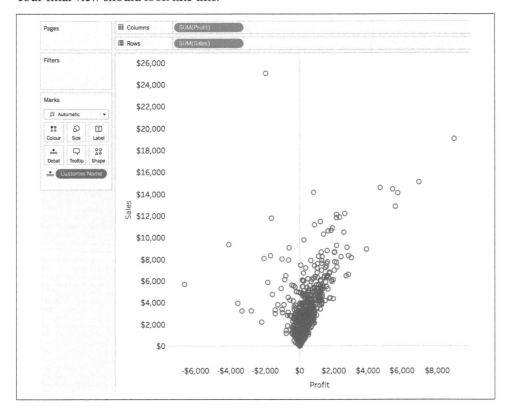

Discussion

A scatter plot is good for comparing two measures. Each mark on the scatter plot allows you to see the sales and profit per individual customer. Holistically, a scatter plot allows you to compare both measures at the same time to see whether a relationship exists between the customer's sales and profits.

You can break down this chart in multiple ways. If you want to add a different shape by category, for example, you can add Category to the Shape property (1) and then click to edit the shapes (2). You can select each category to assign the shapes:

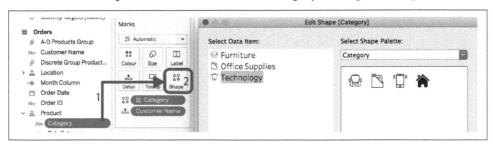

This example uses custom shapes, which is not the default in Tableau, but I'll show you how to get there in the next section.

11.2 Using Custom Shapes

As mentioned in Recipe 11.1, custom shapes are not the default in Tableau, but you can add shapes to your Tableau Repository.

Problem

You want to use custom icons for categories in your business.

Solution

1. Locate your My Tableau Repository folder. This should be within your Documents folder:

2. Inside this folder, you will find the Shapes folder. You can add as many shapes folders as you would like:

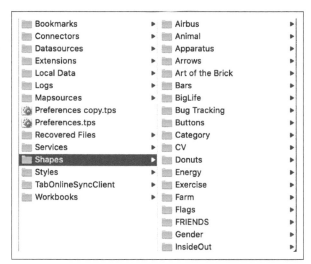

3. Once you have added a new folder with your shapes, you need to reload the shapes into the workbook. To do this, click Shape on the Marks card and select Reload Shapes:

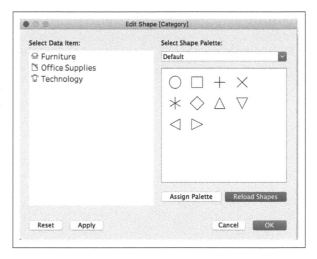

Discussion

Using custom shapes really helps to add to the personalization of your visualizations.

When adding images to your Tableau Repository, I highly recommend creating your own folder or several folders to hold your images. You can add any size shape to your Tableau Repository, anything from your company icons to images you created yourself. The images will be better if they are PNGs without any background.

11.3 Density Marks

Density marks were introduced to Tableau in 2018. Density marks can be used in scatter plots or mapping (Chapter 18). They help to see the concentration of marks in a certain area. The chart we built in Recipe 11.1 shows sales and profit by customers, but a lot of marks are stacked on top of each other near the zero profit mark. This Density mark type allows you to see this concentration more easily.

Problem

You cannot really see or understand the number of marks around the zero profit mark.

Solution

1. Duplicate Recipe 11.1 or re-create a scatter plot.

2. On the Mark Type drop-down, select Density:

Your view should now look like this:

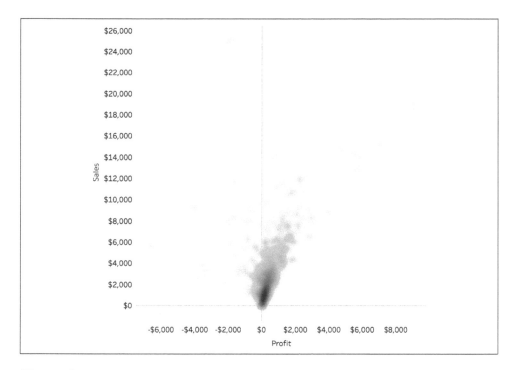

Discussion

The Density mark shows where the highest levels of concentration are within your scatter plot. In this particular example, you can see there are higher numbers between $0 and $2,000 sales values.

You can also change the colors and intensity of the Density marks. If you click Color, you will see the following options:

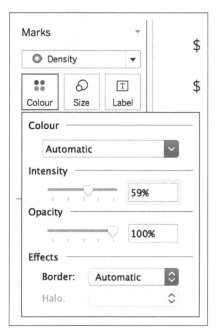

Tableau has specific density color palettes, which appear only when the Density mark type is selected, created specifically to highlight the concentration more effectively:

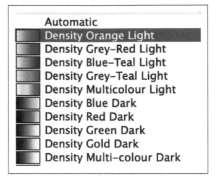

The Intensity slider shows how bright you want the color changes. If you take that up to 100%, this is what it looks like:

Finding the right balance of intensity and opacity is important when trying to make a visualization easy to read.

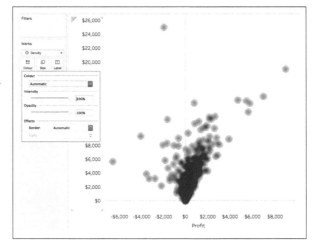

11.4 Clustering

When looking at scatter plots, you might not initially see any groups of marks that are similar to each other. *Clustering* is a way to identify marks that are similar to each other with the ability to add multiple measures that are not in your view.

Problem

You want to compare a customer's sales and profits to group them automatically to find target groups with high profit.

Solution

1. Duplicate Recipe 11.1 or re-create a scatter plot.

2. Go to the Analytics pane inside the worksheet and double-click Cluster:

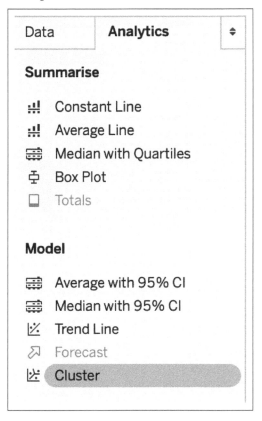

3. This automatically brings up a pop-up box. Here you can add more variables to the cluster analysis and tell Tableau how many clusters you want to create. You can see in the background that Tableau thinks two clusters is what you should be looking for. But we're going to find four clusters:

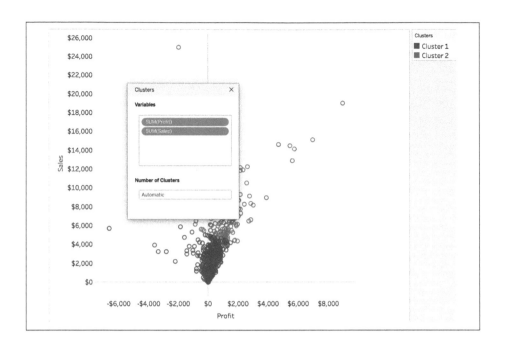

Your new view should look like this:

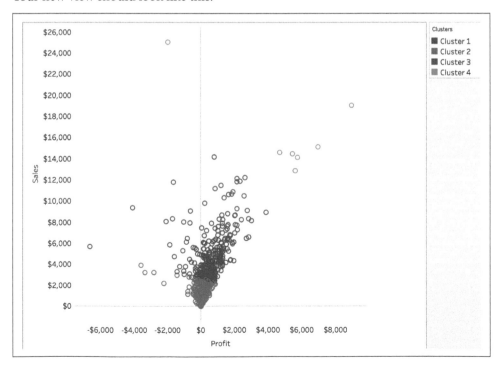

Discussion

Clusters are created using the k-means algorithm. This partitions the data into k clusters. Within each cluster is a center point, which is the mean of all the values within the cluster. K-means identifies the center through a process that reduces the distance between the individual marks and the center point.

Now that we have created the initial cluster, we can do a couple of things to enhance it.

If you want to rename the clusters to be more specific, you have to save the cluster as a group. You can do this by dragging and dropping the cluster from Color on the Marks card to the Dimensions pane:

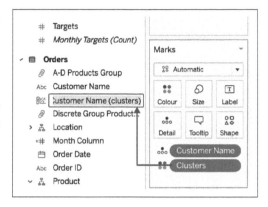

You can now right-click "Customer Name (clusters)" to edit the group. This will allow you to change the name of the clusters:

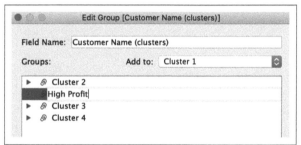

However, now that you have saved the cluster, it is no longer dynamic to the view. You have to right-click the cluster and select Refit to change the clusters based on new measures or marks. This will also default the clusters back to the number of automatic clusters.

You can also find out more information about the clusters by right-clicking the cluster and selecting Describe Cluster, which will give you the following information about the summary and models:

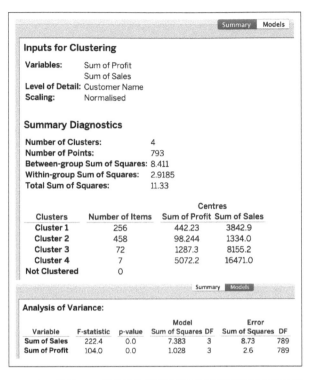

Summary Models

Inputs for Clustering

Variables:	Sum of Profit
	Sum of Sales
Level of Detail:	Customer Name
Scaling:	Normalised

Summary Diagnostics

Number of Clusters:	4
Number of Points:	793
Between-group Sum of Squares:	8.411
Within-group Sum of Squares:	2.9185
Total Sum of Squares:	11.33

		Centres	
Clusters	Number of Items	Sum of Profit	Sum of Sales
Cluster 1	256	442.23	3842.9
Cluster 2	458	98.244	1334.0
Cluster 3	72	1287.3	8155.2
Cluster 4	7	5072.2	16471.0
Not Clustered	0		

Analysis of Variance:

Variable	F-statistic	p-value	Model Sum of Squares	DF	Error Sum of Squares	DF
Sum of Sales	222.4	0.0	7.383	3	8.73	789
Sum of Profit	104.0	0.0	1.028	3	2.6	789

If you have saved the cluster to reuse in different views and your data changes, you can right-click the cluster and select Refit:

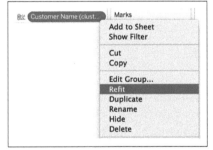

This will recompute the clusters according to the refreshed data. However, when you refit a cluster, new clusters will be created, and they will replace any existing aliases of your previous clusters.

11.5 Trend Lines

Trend lines are used to identify trends in your data. These can be used on line charts and scatter plots. When using them on scatter plots, you will see the significance of the relationship between the two measures in the view.

Problem

You want to see the trends between sales and profit for each category, split by subcategories and product name.

Solution

1. Create a new scatter plot with Sales on Columns and Profit on Rows. You should add Category to Color and then drill down to Product Name on the Marks card:

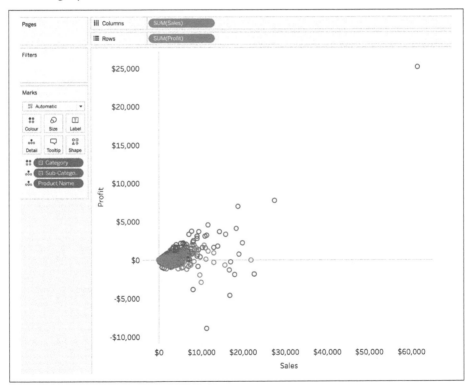

2. In the Analytics pane, double-click Trend Line to add it to the view:

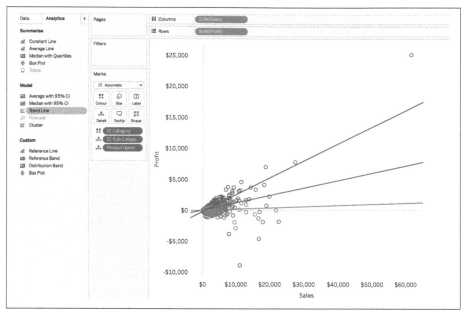

Now you have a linear trend line per category.

Discussion

Using trend lines, you can see that the Technology category has a higher slope, meaning more sales equals higher profit. This is one use case of trend lines, and it uses the Linear option. To edit this type of trend line, right-click any of the trend lines and select Edit All Trend Lines:

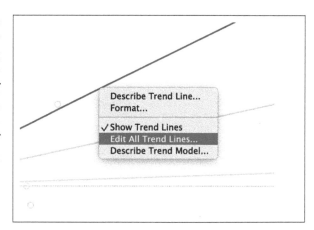

In the pop-up box you have the option to change the model type. The model type will be dependent on your data. Use the other model types only if one fits the data better.

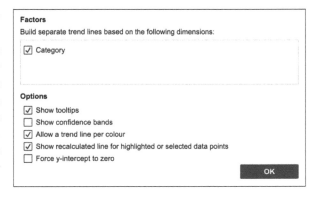

You also have the option to turn off a trend line per category (and color) and just have it as an overall trend line for all the marks. You also have different options to select. The tooltip shows the model of the trend line and includes its significance. Confidence bands show 95% upper and lower limits of the model.

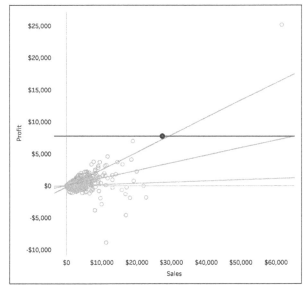

The "Show recalculated line for highlighted or selected data points" is useful but can also be a nuisance. If you select a group of marks, the trend line will be recalculated, which could be what you are trying to do, but if you select a single mark, it will still recalculate the trend line, but it will just be flat, as shown here:

Adding Drop Lines

With any mark, you have the option to add drop lines. Drop lines add a line to both the x- and y-axes to show specifically what the values are. To turn them on, right-click any mark in the view and choose Drop Lines > Show Drop Lines:

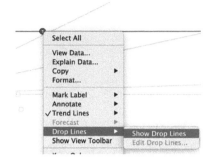

This shows a gray drop line to both axes:

To add labels to your drop lines, which allow them to be read easier, you need to right-click any mark and choose Drop Lines > Edit Drop Lines:

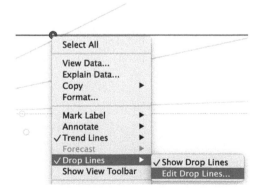

You can choose which axis you want the drop lines to apply to, whether you want them to appear when selected or to be always on, and finally, whether you want the labels to appear. I have turned the Labels to Automatic:

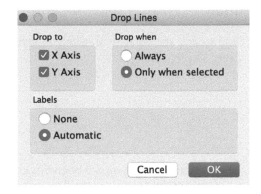

This now adds a label to the drop lines when you have selected a single mark:

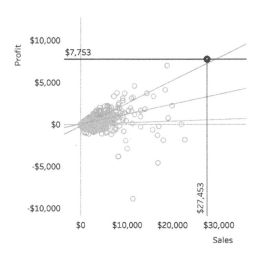

Finally, the last option in Edit All Trend Lines is to force the y-intercept to be zero. This option should be checked only if you have statistical reasoning for it.

11.6 Explain Data

Those with a keen eye may have noticed the outlier in our scatter plot that is way above all the other sales and profit values. Tableau has a quick way to investigate this mark, and this is called Explain Data.

Problem

You want to investigate the possible reasons a data point is such an outlier by using Explain Data.

Solution

1. Right-click the mark you want to investigate and select Explain Data:

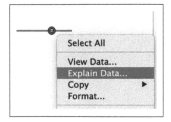

2. This will bring up a pop-up box with explanations as to why this point is an outlier, using both measures in the view. Read these explanations to learn more about the mark.

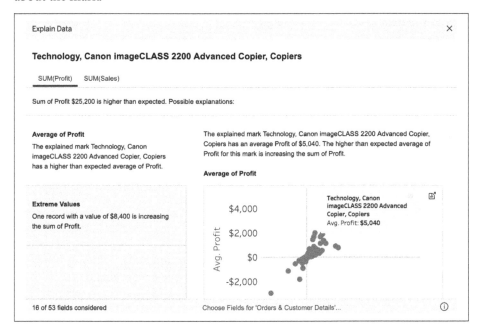

Discussion

The Explain Data pop-up has a tab per measure within your view with possible explanations of why this mark is so different from the other marks in the view. In this example, both tabs explain that this particular mark has a higher-than-average sales and profit value, and this is just a single record that is increasing both the sales and profit values. You can, therefore, decide whether you want to remove this product from this view or keep it.

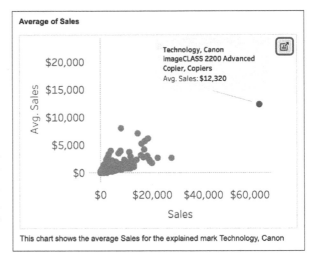

Inside the Explain Data pop-up, you also have the option to export the charts to keep them as sheets in your workbook. You can do this by clicking the top-right corner of the chart:

11.7 Connected Scatter Plot

A *connected scatter plot* allows you to track how a particular item has performed over time for two measures.

Problem

You want to see how each category has changed over time for both sales and profit.

Solution

1. Create a scatter plot with Profit on Rows, Sales on Columns, and Category for Color and Shape:

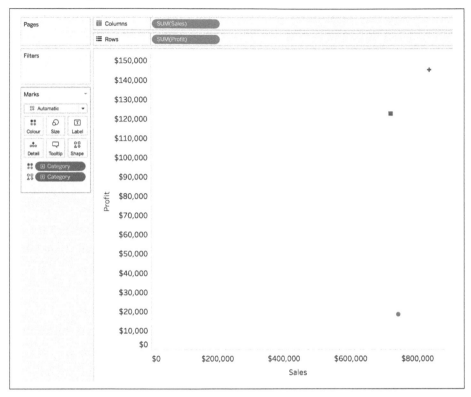

2. Add continuous months to Detail on the Marks card:

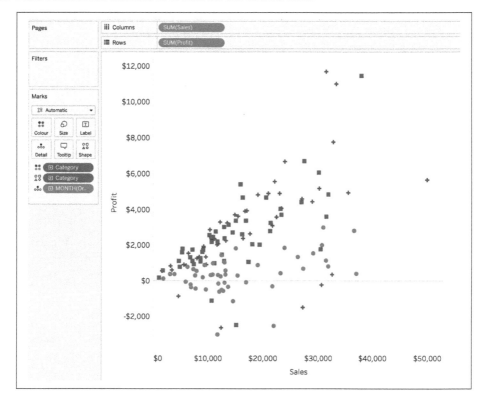

3. To simplify the view, filter to the last year in the data.

4. To create the connected scatter plot, I recommend using a dual axis, which allows you to have the circle or shape for each mark along with the line. Press Ctrl (Command on Mac) and duplicate *either* Sales or Profit onto the Rows or Columns shelf:

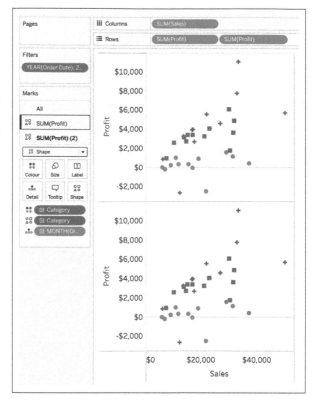

5. On the second Marks card, change the type to Line and add continuous months to Path:

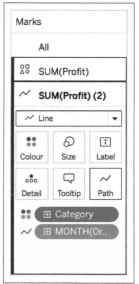

Your view will now look like this:

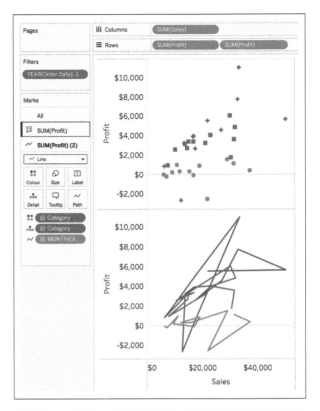

6. Finally, you need to make this into a dual axis to end up with a view like this. And don't forget to synchronize:

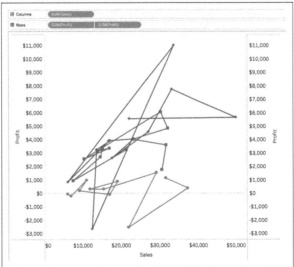

Discussion

A connected scatter plot allows you to see how dimensions change over time. Each line represents a category in our example. If we look at Furniture, we can see that sales are increasing over time (profit on the whole is above $0), but in October 2021 sales dip to –2,500:

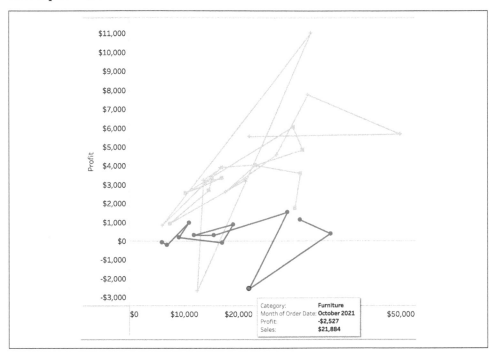

The connected scatter plot shows the relationship between two measures.

If you add MONTH to the Pages shelf, you will get a set of controls that allows you to play the animation over time. I recommend checking the "Show history" box, changing "Marks to show history for" to be All, and choosing Show Both (Marks and Trails):

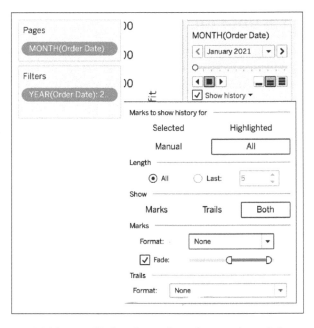

Now when you press the right arrow, Tableau will play through each month and show the progression of sales and profit over time:

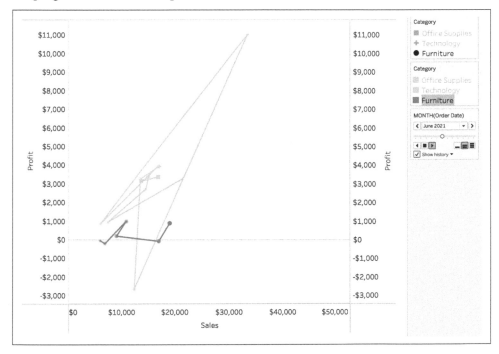

11.8 Dot Plot

Keeping with the circular theme in this chapter, we move on to *dot plots*. These are a simple way of showing a change or a range of data across multiple dimensions. These can also be known as *barbell*, *dumbbell*, or *DNA charts*, especially when comparing two dimensions.

Problem

You want to visualize the range of sales values: minimum, maximum, and average, by subcategory.

Solution

1. Add Sub-Category to Rows.

2. Right-click and drag Sales to Columns; then select AVG(Sales):

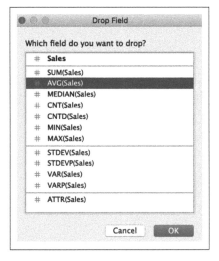

3. To get the different circles per measure, we need to have a shared axis. To do this, right-click and drag Sales on top of the average sales axis and then select MIN(Sales):

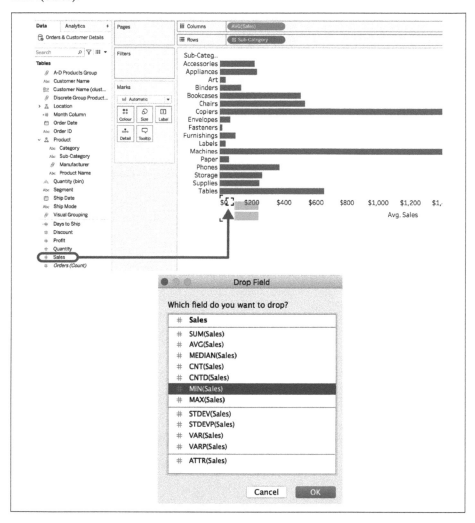

By doing this, we have created a Measure Names and Measure Values chart:

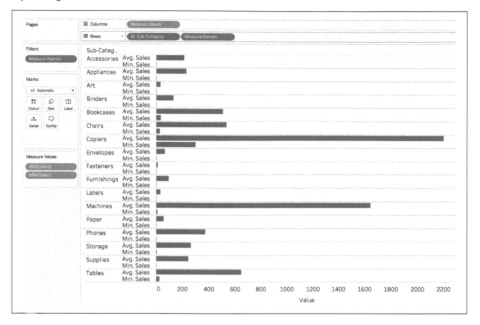

4. We need to add one more measure to the Measure Values shelf—maximum sales. Right-click and drag Sales to the Measure Values shelf and select MAX(Sales):

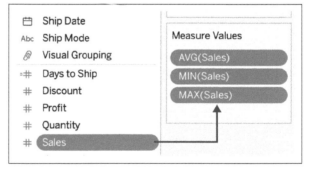

5. From here, we need to move Measure Names from Rows to Color on the Marks card:

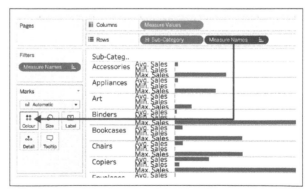

6. Change the mark type to Circle or Shape:

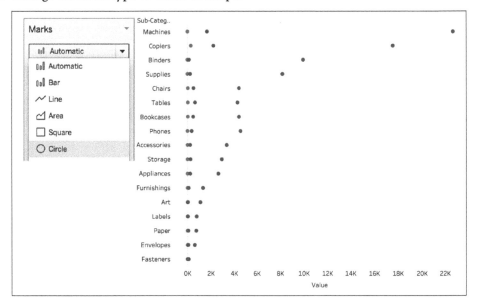

You can now see the range of sales for each subcategory (this view is sorted by sum of sales):

7. One last thing I like to do on these charts is add a faint gray line to the background to truly highlight the range between the values, which is converting it from a dot plot into a barbell chart.

   To do this, duplicate (press Ctrl or Command) Measure Values field on Columns:

8. Right-click the second Measure Values and select Dual Axis:

9. Always synchronize your axes. Right-click an axis and click Synchronize.

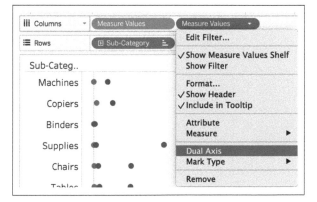

10. Right-click the second Measure Values field and change the mark type to Line:

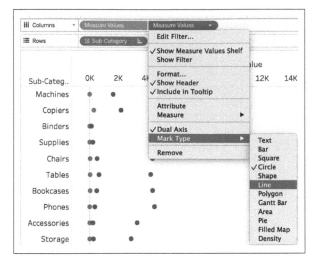

11. We need Tableau to draw by the Measure Names. On the second Marks card, move Measure Names from Color to Path:

12. Finally, change the color of your line, make it thinner, and move marks to the back—voilà, a dot plot:

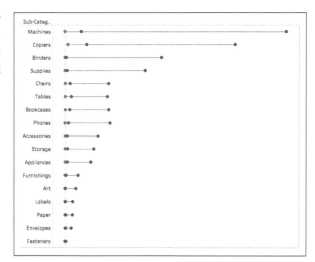

Switching Axes

You can move your marks to the front or back in two ways. The marks at the front are usually the second field in your dual axis. You can right-click the axis and select "Move marks to back/front" (depending on which axis you have clicked), or you can rearrange the measures on Rows or Columns:

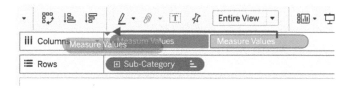

Discussion

Dot plots are a simple way of showing ranges within your data. This could be minimums/maximums, or information like the gender gap, or last year versus this year.

One thing to be aware of is that the tooltip for the line could be misleading, and therefore, you should consider turning off the tooltip for the line.

11.9 Dot Strip Plot

Similar to the dot plot, a *dot strip plot* shows the range between all the values. It allows you to see whether a higher concentration of marks is toward one extreme or the other and whether they are evenly distributed.

Problem

You want to see the distribution of sales for all products within each subcategory.

Solution

1. Add Sub-Category to Rows and Sales to Columns:

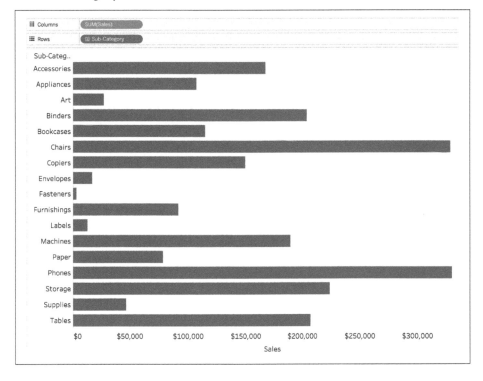

2. Change the mark type to Circle:

3. We now want to break down this view down by product name. Do this by adding Product Name to Detail:

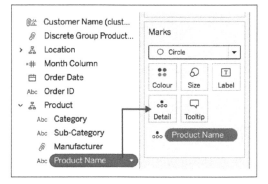

4. With so many products overlapping, it is recommended that you reduce the opacity of the marks to enable the overlapping products to be partially visible. To do this, select Color on the Marks card and move the Opacity slider to a lower number:

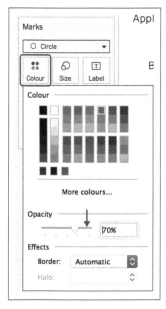

Your final view should look like this:

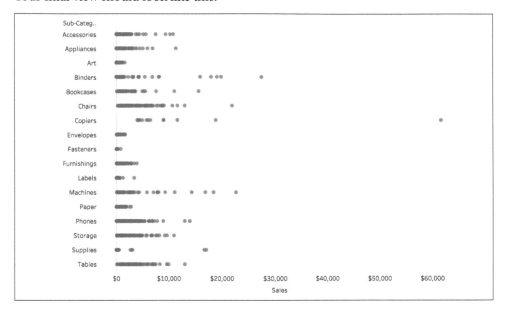

Discussion

The dot strip plot allows you to see the full range of values for each product, allowing the audience to see outliers. The downside to having all the similar sales products stacked on top of one another is that you can't see every single product individually. Being able to see each product could be a requirement, in which case, you could create a jitter plot.

11.10 Jitter Plot

A *jitter plot* gives each product a unique value, to spread the products across an axis.

Problem

You want to see each product's sales per subcategory.

Solution

1. Duplicate Recipe 11.9 or re-create a dot strip plot.
2. You could use one of two calculations for this, but they are dependent on your data sources and requirements (see Discussion).

3. Create a calculation with the following syntax:

 RANDOM is an undocumented function, but it assigns every mark in the view a random number between 0 and 1.

4. Add this calculation to the Rows shelf and deselect Show Header:

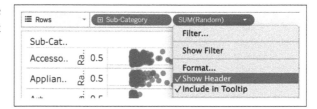

5. I recommend reducing the opacity and adding a border to each mark:

Your final view should look like this:

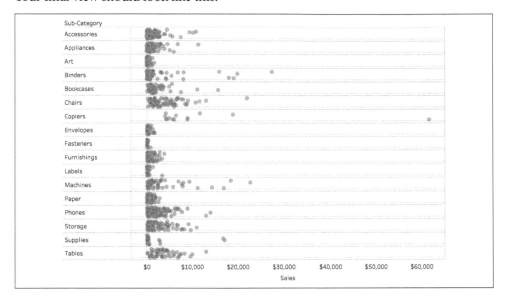

Discussion

As mentioned, the jitter plot shows the spread of sales by individual products, and shows the concentration a little bit better. The RANDOM calculation gives each product a random number between 0 and 1 and will appear in the tooltip. This might confuse consumers. To remove it from the tooltip without editing the whole tooltip, right-click the Random field in Rows and deselect "Include in Tooltip":

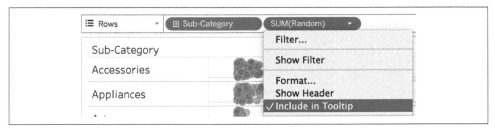

To make sure the y-axis is not misinterpreted for a different variable, make sure you give your users information about why the marks are spread out.

In this recipe, I mentioned the possibility of using one of two calculations. The RANDOM calculation is undocumented and unsupported for some data sources. If this does become an issue, you can still create this chart, but instead of using RANDOM, you will use Index.

INDEX is a table calculation that gives each mark a sequential number from 1 to X. To do this chart with INDEX instead, create a new calculation with the following formula:

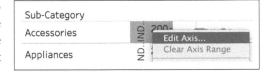

We add this calculation to Rows in place of the RANDOM calculation. You should then change the way your table calculation is computing; we want to restart the numbering per subcategory; therefore, we need Product Name selected:

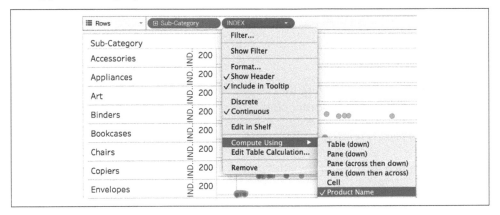

There is an extra step here, which is to make each y-axis independent of the others. Start by right-clicking the Index axis within the view and select Edit Axis:

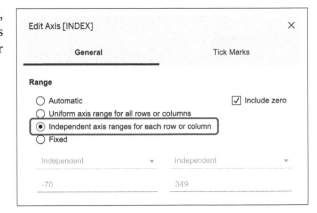

Then, in the pop-up box, select "Independent axis ranges for each row or column":

Finally, repeat step 5 from the recipe. Right-click the INDEX field in Rows and deselect Show Header and Include in Tooltip:

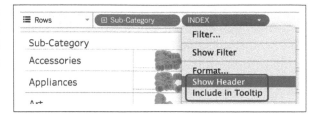

There are no performance differences between using RANDOM and INDEX; it comes down to personal preference and which data source you are using.

11.11 Box Plots

Box plots are known for showing the distribution of data along an axis. These can be used to enhance dot strip and jitter plots.

Problem

You want to add a box plot to the jitter plot to have statistics on the range of data per subcategory.

Solution

1. Duplicate Recipe 11.10 or re-create a jitter plot.

2. In the Analytics pane, drag and drop Box Plot to the SUM(Sales) box on the view:

The whiskers of the box are currently set to "Data within 1.5 times the IQR" of the data:

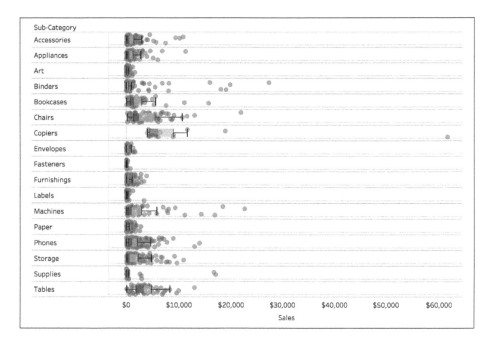

 IQR (interquartile range) is where the majority of values are situated.

3. This can be changed to show the maximum extent of the data. Right-click any box plot in the view and select Edit:

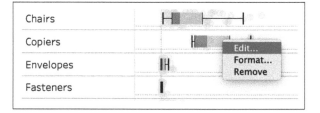

4. Use the drop-down to change the whiskers to "Maximum extent of the data":

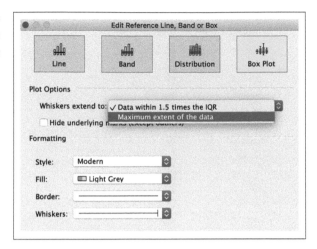

Your final view should look like this:

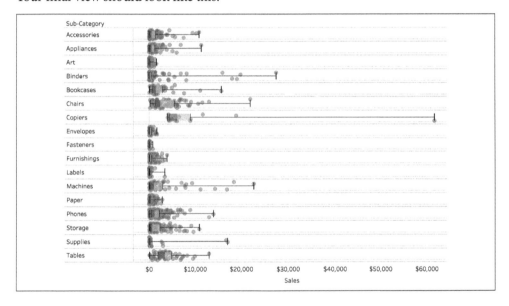

Discussion

A box plot allows you to add information about the range and spread of the data. When you hover over a box plot, Tableau gives you the following information, which tells you how far the data is spread:

When you click to edit a box plot, you'll see a variety of formatting options, including colors and whisker styles:

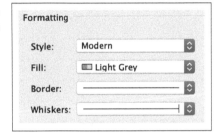

Box plots also appear under the Show Me menu, which allows you to quickly create a box plot chart (without the jitter). To create a box plot using the Show Me menu, you will need to select which fields you want to use before selecting the box plot from Show Me:

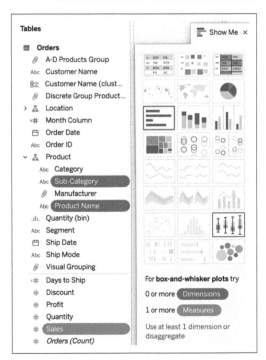

This will create a vertical axis box plot:

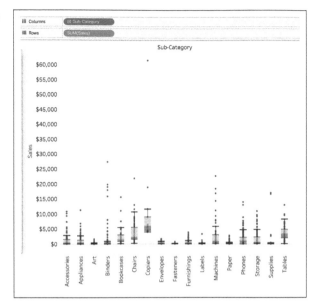

To quickly switch the axis for a horizontal axis, click the Swap Axis button on the toolbar:

One other way you can add a box plot to your view is by right-clicking the axis and selecting Add Reference Line. In the pop-up box, you have the option to select Box Plot:

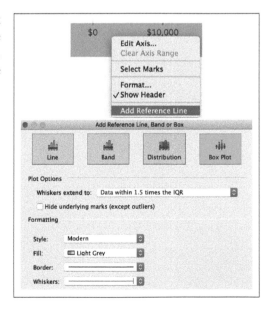

11.12 Using Shapes for BANs

BANs (Big Actual Numbers) are discussed in Recipe 4.5. BANs can be text-based, and to add extra context, you can use shapes to show whether the change is positive, negative, or neutral.

Problem

You want to use shapes to find out whether the latest month of data for sales represents an increase (upward triangle) or decrease (downward triangle) when compared to the previous month's sales.

Solution

1. Filter to the latest year of data and add Month of Order Date to Rows.

2. Change the mark type to Shape:

3. Add the calculation Sign Percent Difference Sales to Shape on the Marks card and make sure Compute Using is set to Order Date (this calculation was created in Recipe 4.10):

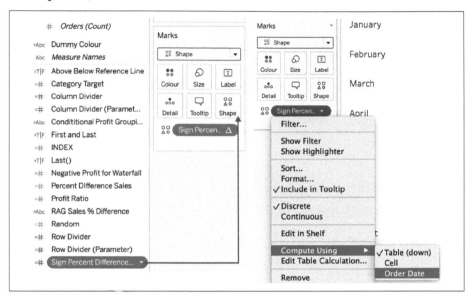

4. Click Shape on the Marks card to change the shapes accordingly:

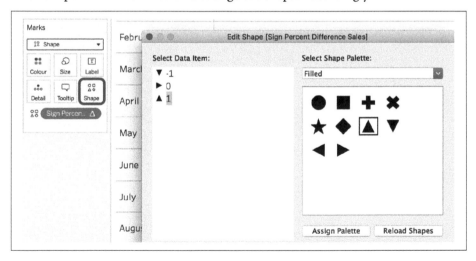

Your view should now look like this:

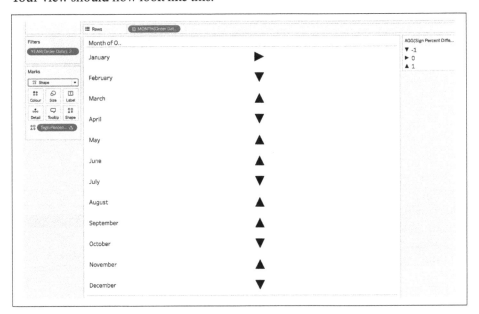

5. Add the previously created calculation of Last (also created in Recipe 4.10) to Rows (1) and hide the false rows (2):

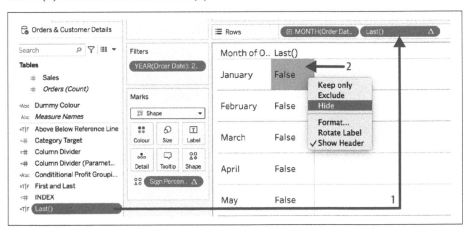

This is now the view:

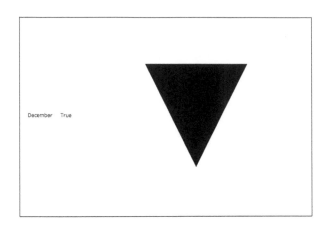

6. Finally, clean up the view by hiding the headers, adding values to Text on the Marks card:

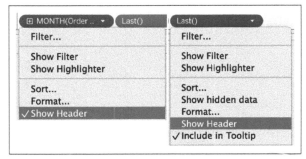

7. Also, duplicate the Sign calculation to Color and add Sales and Percent Difference Sales to Label:

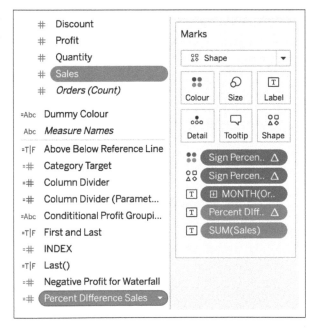

8. Depending on the size of your shape and the space it needs on the dashboard, you might need to increase the font size. You can do this by clicking Label on the Marks card, and selecting Font to change the size:

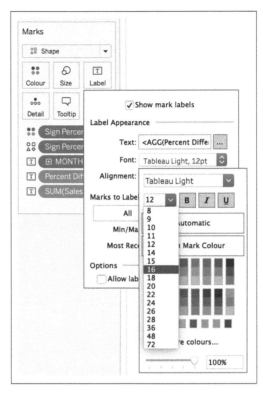

Your final view should look like this:

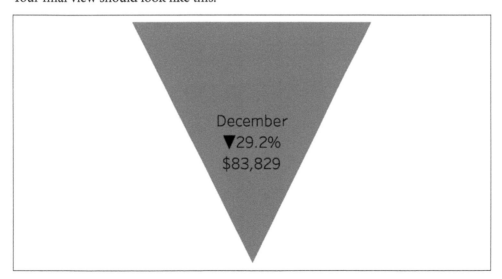

Discussion

Using shapes for BANs and KPIs gives a quick visual explanation as to whether the value is above or below the previous value.

See Also

Adam McCann's KPI reference for options on how to format your KPIs and BANs (*https://oreil.ly/y8Lrh*)

11.13 Pie Charts

Pie charts are taught at such an early age, so everyone is likely to know how to interpret them. They represent a part-to-whole relationship and are relatively easy to create in Tableau.

Problem

You want to compare the percentage of female shoppers to males.

Solution

1. Add Gender to Color on the Marks card:

2. Change the mark type to Pie:

3. Right-click and drag Customer ID to Angle on the Marks card, and select CNTD:

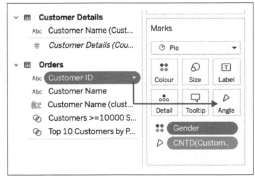

This creates the following view:

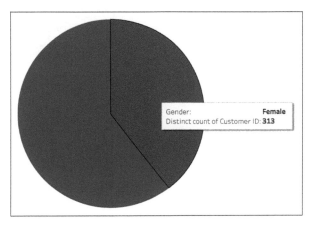

4. To add the percent of total customers, duplicate CNTD(Customer ID) onto Label:

5. Add a Quick Table Calculation of Percent of Total onto the second CNTD(Customer) field on the Marks card:

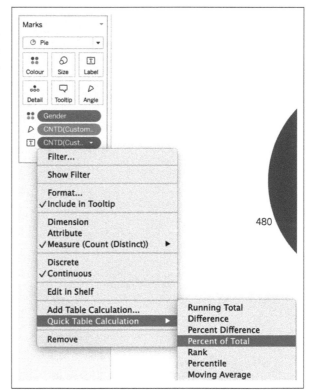

Your final view will look like this:

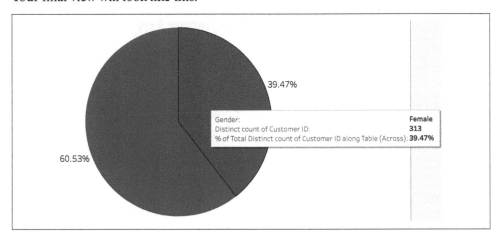

Discussion

Pie charts work really well when comparing a maximum of three to four segments, or when actual values are not important (versus getting a rough idea). In this case, gender works well, but if you want to see a pie chart with all the subcategories, I advise against this because other chart types can show you that information in a better way.

11.14 Donut Chart

Sticking with food-themed charts, similar to the pie chart, a *donut chart* also compares part-to-whole relationships, but this time with a hole in the middle. Donut charts have a better visual appeal compared to pie charts, and they allow you to add information into the middle of the chart.

Problem

You want to create a donut chart showing a male-to-female shopping split, but with the center gap showing you just the female percentage.

Solution

1. Duplicate Recipe 11.13 or re-create a pie chart.

2. On the Columns shelf, double-click and type **AVG(0)**:

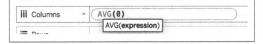

3. Press Ctrl and duplicate the AVG(0) field into Columns. You should now have two AVG(0) fields on Columns:

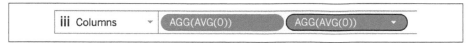

4. On the second Marks card, change the type to Circle:

5. Move Gender from Color to Detail and change the color of the circle to match your background color:

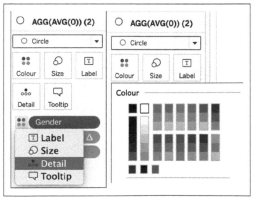

6. Change the alignment to Center, select Min/Max on the Marks to Label section, and uncheck "Label maximum value" at the bottom. This is because the female is the lowest value, so the label selection is dependent on your data:

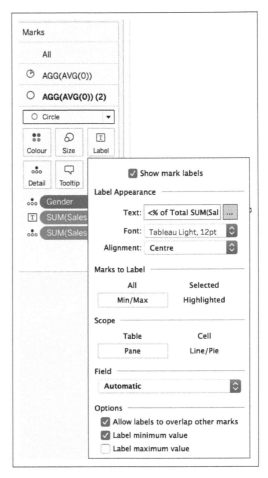

7. On the All Marks card, change the overall size to the midpoint:

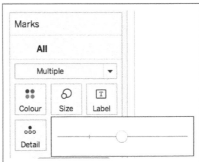

8. Now we need to create a dual axis. To do this, right-click the second axis and select Dual Axis:

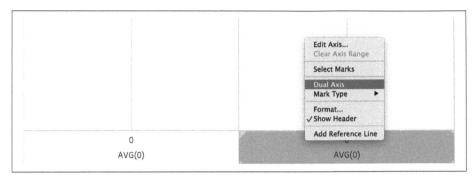

Your view will now look like this:

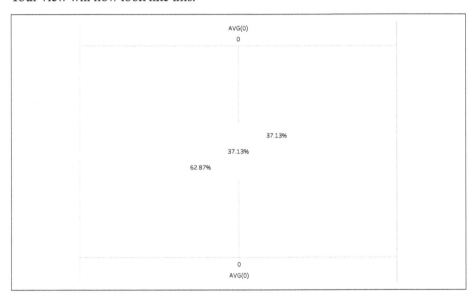

9. Don't panic, we haven't broken the view. You need to go back to the second Marks card and adjust Size to smaller. The size you make this is dependent on the style you want to go with:

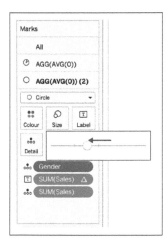

You now have created your donut, and it should look like this:

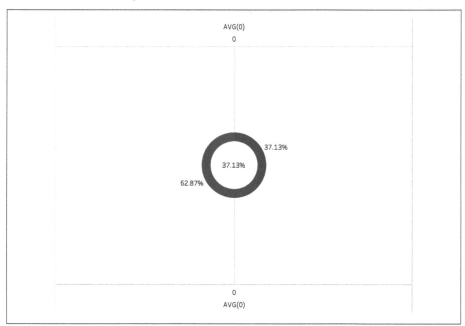

10. The last step is to tidy the view, hide the headers, and remove the zero lines:

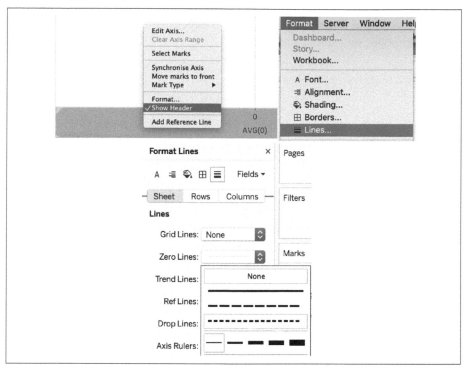

Your final donut chart should look like this:

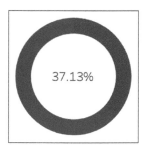

Discussion

Donut charts have a better visual appeal on dashboards, especially when using them to indicate how far along a project is. You can use a donut chart for any use case where the total is 100%. However, a drawback arises when you want to measure progress to a goal: you cannot tell how far past 100% you have achieved. This is where a bullet chart would be recommended (Recipe 3.9).

The style you want determines how thin or thick the donut is; there are no strict rules.

Summary

Throughout this chapter, you have learned how to use three mark types: Shape, Circle, and Pie. All of these can be used to enhance your visualizations, enabling you to have multiple dimensions on the Marks card.

Being able to create scatter plots and pie charts is fundamental to understanding your data.

CHAPTER 12

Gantt

A *Gantt* is another mark type that can be used in Tableau. Gantt charts can be used to show duration, activities, and highs/lows.

12.1 Basic Gantt Charts

A basic Gantt chart uses the duration of an event or activity and adds it to Size. This allows you to see how long a particular event or activity lasted.

Problem

You want to show the time span between the order date and ship date in days for every order, sorted by the longest shipping time.

Solution

1. Add Order ID to rows:

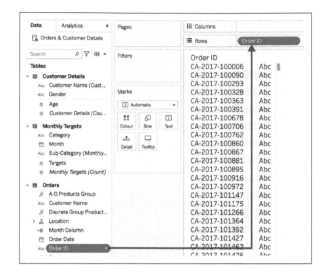

2. Right-click (Option-click on Mac) and drag Order Date to Rows, then select Discrete date:

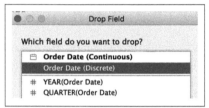

3. Right-click and drag Ship Date to Rows; then select Discrete date:

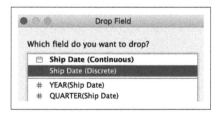

Your view should look like this:

4. Add continuous Order Date to Columns:

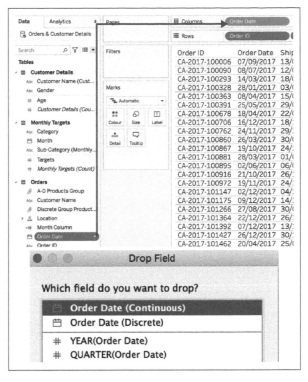

5. Next, we need to calculate the number of days between the order date and ship date. This can be done by creating the following calculated field:

```
Days to Ship
```

```
DATEDIFF('day',[Order Date],[Ship Date])
```

6. Once you have this calculation, add it to Size on the Marks card:

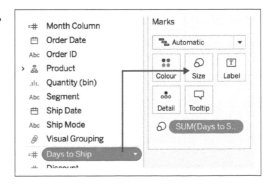

7. Finally, I like to sort by the days to ship. You can do this by right-clicking Order ID and selecting Sort:

8. On the Sort By option, select Field, and then Descending. Days to Ship was already prepopulated:

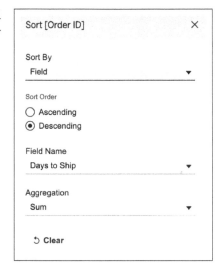

Your final view should look like this:

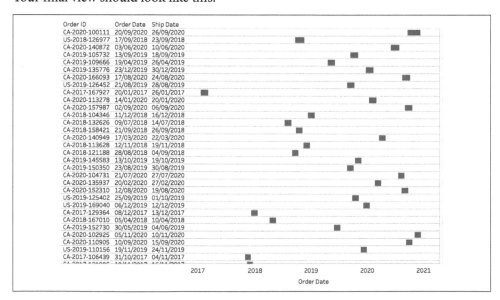

Discussion

A Gantt chart is very good at showing duration and timing. Using the Days to Ship calculation, you can see which orders took the longest to ship. Having this type of calculation could also be used for project management or whenever you have a start and end date.

12.2 Barcode Charts

Gantt has many use cases, and one of those is to build a barcode chart. *Barcodes* are used to see data and can be read by a machine. In Tableau, they can be represented by various widths or heights and can show a lot of information in a small amount of space.

Problem

You want to see how each region's profit changes over time.

Solution

1. Start by adding continuous Order Date to Columns:

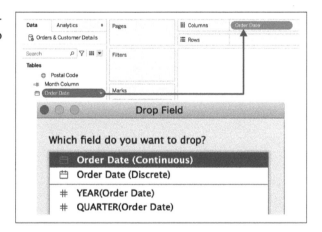

2. Add region to Rows:

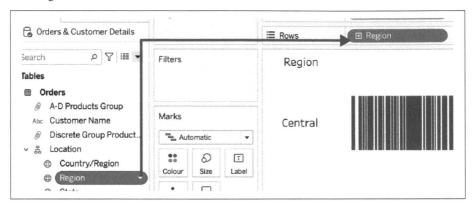

3. Notice that Tableau has defaulted to a Gantt Bar mark type:

4. The final step is to add a calculation to color. We can use a previous function called SIGN. Create the following calculation:

5. We will need to convert this calculation to discrete to allow for discrete colors:

6. Add the calculation to Color on the Marks card and change the colors to suit.

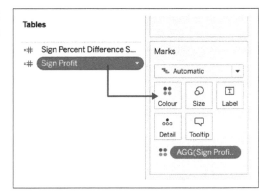

Your final view should look like this:

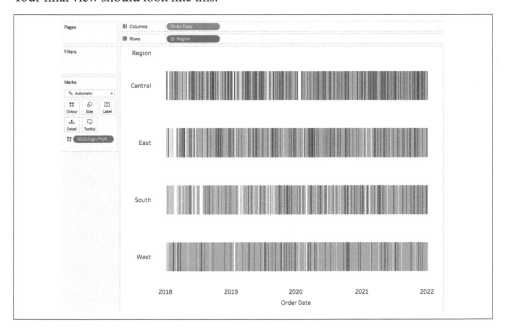

Discussion

Creating a barcode chart allows for quick visual analysis of where the positive and negative profit is on specific days.

Instead of using the function SIGN, you could create a calculation that groups the profit into several categories and use that on Color instead.

12.3 Waterfall Charts

A *waterfall chart* shows the overall running total along with the positives and negatives. Each part of the waterfall chart contributes to the overall outcome and is used to visualize part-to-whole relationships.

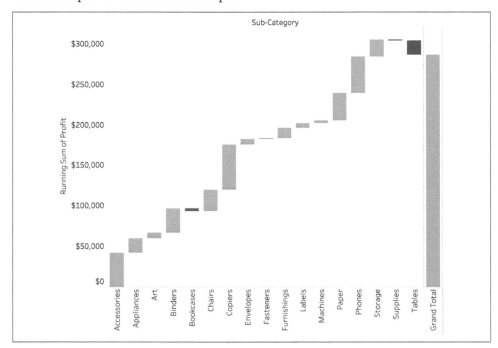

Problem

You want to see the running total of profit and loss over the month.

Solution

1. Filter to a specific month and year in the data. For this example, we want to look in only a particular month; otherwise, you don't need to filter.

2. Right-click, and drag Order Date to Columns and select discrete day:

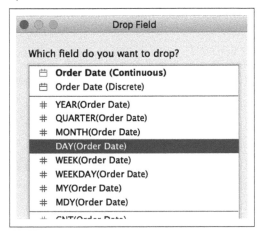

3. Set the view to Entire View:

4. Add Profit to Rows:

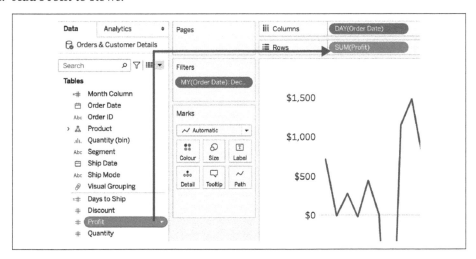

5. To get the running total, add the table calculation to the SUM(Profit) field on Rows:

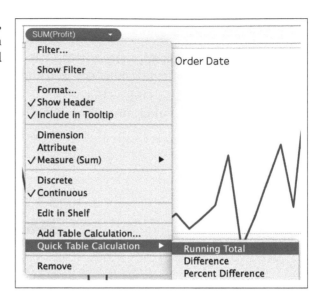

6. Change the mark type from Automatic to Gantt:

The view should look like this:

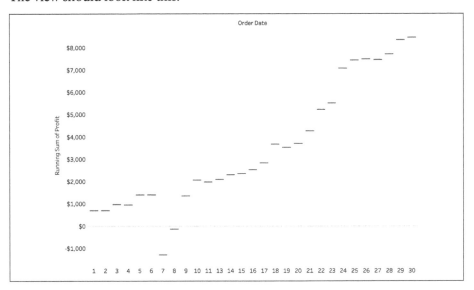

7. Add the change for each day in the month. You will need to create a calculation that uses the negative sum of profit:

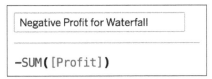

Negative Profit for Waterfall

−SUM([Profit])

8. Add this calculation to Size on the Marks card:

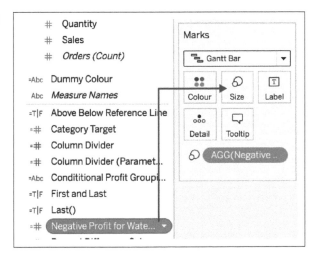

The view should look like
this:

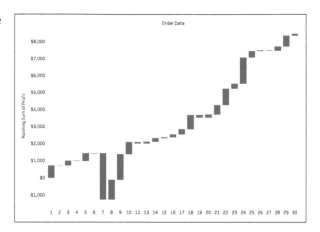

9. The last step is to visually show the profit and loss by using colors. We can use
   the previously created Sign Profit calculation (Recipe 12.2).

10. Add this calculation to
    Color on the Marks card
    and choose appropriate
    colors to represent posi-
    tive and negative:

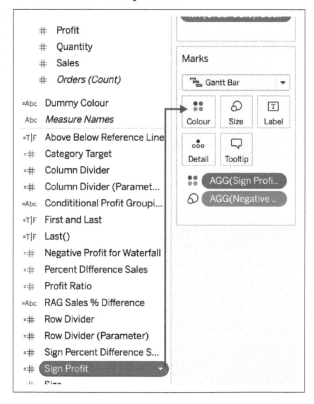

Your final view should look like this:

Discussion

Waterfall charts are good at showing profits and losses over a time period. You can see when your profit is at a loss. Another use case is for breaking down income versus expense, which is usually sorted from largest value to smallest. Then we can add another column for the remaining balance.

It's a good idea to add the grand total or additional column for balance to help users read this chart, which would then show the total profit of this period. To do this, choose Analysis > Totals > Show Row Grand Totals:

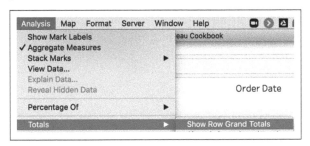

This makes your view look like this:

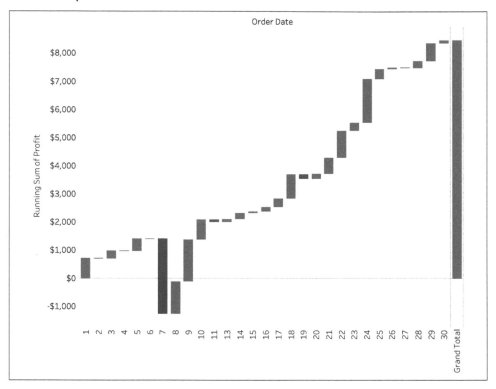

You might also notice that the tooltip shows the reverse profit. This is due to the calculation we needed to create the waterfall. You can change this in two steps. First, right-click the Negative Profit field on the Marks card and deselect "Include in Tooltip":

The second step is adding Sum of Profit to Tooltip on the Marks card:

Now in your tooltip you will have the correct profit value.

Depending on how your data is structured and what questions you want to ask, you might require some data preparation before creating a waterfall. For example, if you have a column per measure that you want to visualize, you can't use Measure Names or

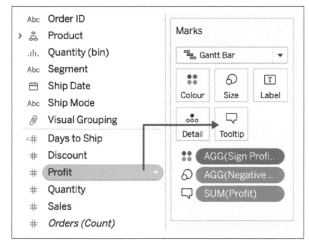

Measure Values to re-create a waterfall in the most efficient way without creating a calculation for every measure. Therefore, your data would need to be pivoted to allow for a row per measure.

See Also

For any data preparation tips and advice, take a look at Carl Allchin's book, *Tableau Prep* (O'Reilly).

For other waterfall use cases and examples, check out Ryan Sleeper's book, *Innovative Tableau* (O'Reilly).

Summary

The Gantt Bar mark type is useful for showing periods, project schedules, or profit and loss as a waterfall chart. You can now build each one of those charts for your projects.

CHAPTER 13

Stories

In Chapter 7, I introduced the concept of bringing together worksheets to make a dashboard, which can be used to examine relationships between the data points. Story points use dashboards or worksheets to create a *story*, or a series of visualizations, to convey a particular message, like a presentation.

With a story, you can guide a user through a journey of analysis without being with them; this enables the user to find the intended story of the data but gives them the ability to find their own story, too. A story can be like an interactive PowerPoint document, where data is front and center of the story, with the ability to interact with the story without it affecting the next point. In this chapter, we will be covering how to create different elements of story points in Tableau.

13.1 Creating a Story

Problem

You want to start creating a story to guide a user through a piece of analysis.

Solution

1. To create a story, click the button at the bottom on the far right, which looks like a book with a plus sign:

You can also create a new story by using the toolbar at the top to select Story > New Story:

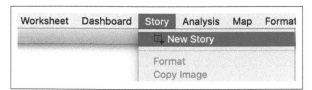

This gives you a blank canvas, known as a *story point*:

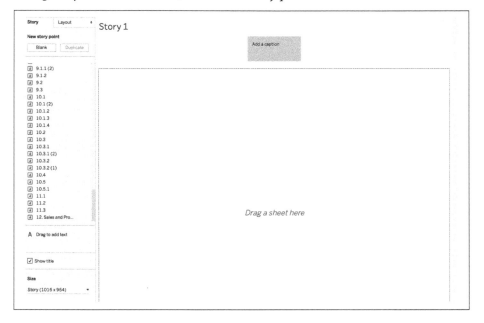

2. From here, you can add a worksheet or dashboard into a single story point by dragging and dropping them into the middle of the canvas:

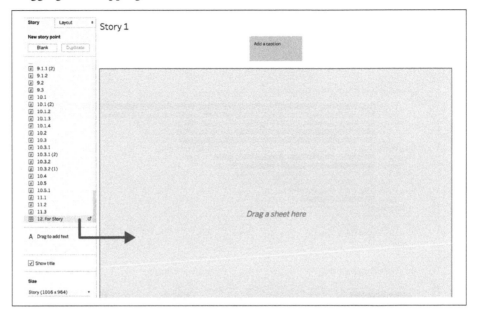

 You can add only a singular worksheet or dashboard into a story point. If you want to add multiple worksheets, you should create a dashboard with the worksheets on, and then add that to the story point.

When adding the dashboard to the canvas, it should look like this:

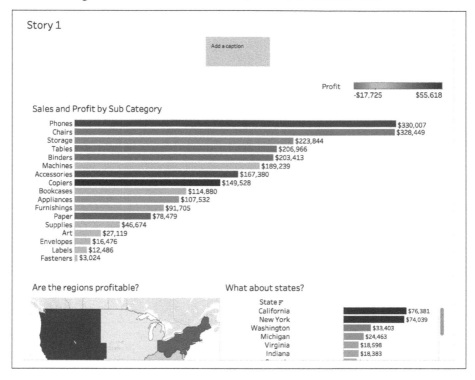

The dashboard used in this story can be re-created using examples throughout this book.

3. Edit the navigation caption to tell your audience to **Start Here**:

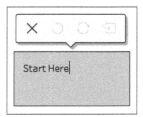

Discussion

A story can be used with either dashboards or sheets. You cannot create a dashboard within a story, so it is advisable to create the dashboards before starting your story. A story can be interactive as it guides the reader through the journey by using the captions at the top, accompanied by forward and backward arrows. These can be changed and are covered later in this recipe. When you create a story and use the interactivity, it is important to note that this won't be reflected within the standalone dashboard or worksheets.

The story workspace has some similarities and differences compared to the dashboard workspace. There are different controls, formatting options, and some limitations when creating a story. Each section will be discussed within this chapter.

Before really getting started with creating your story, it is important to format the different elements within the story point, to allow for the dashboard or worksheets to fit the stories.

I recommend aiming to fit a story point onto a single screen rather than having scroll bars.

To change the size of the story, make sure you are on the Story tab on the left, and go down to the size option at the bottom. It has a range of different sizes to select from; pick one suitable to your screen size:

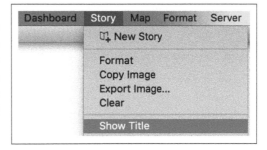

The next important step is thinking about whether your story needs its own title or if you will to use dashboard/worksheet titles to narrate your story. If you want to turn off the title to allow for more space, you can do this by going to Story in the toolbar and unchecking Show Title:

The default formatting for the navigator style for each story point is by using captions, which allows you to add any free text into this box. The different navigator styles are a personal preference and depend on the story you are trying to tell. To change the navigator style, which again could allow more space for the dashboard inside the story, you need to click the Layout tab on the left, where you have different options for the navigator style:

Numbers shows each story point as a different number:

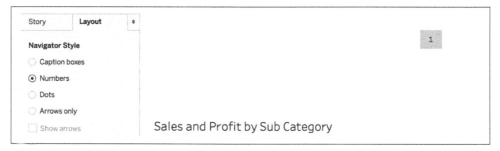

Dots uses breadcrumbs to show your user there are multiple story points:

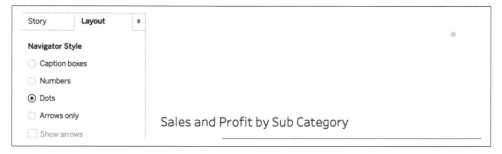

Finally, "Arrows only" shows you which story point you are on compared to the total number of story points in the story.

The "Show arrows" option will be available when you have multiple story points.

Also, by default, the back-ground color of the navigator style is gray for captions, numbers, and arrows. You can change this and many global story settings by choosing Story > Format in the toolbar at the top:

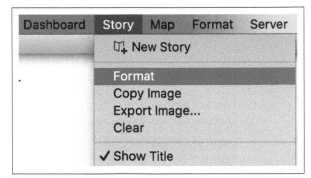

This brings up a toolbar on the left side with a variety of options, one of which means you can change the navigator shading:

Once you have made all the changes to the story formatting, you might notice that your dashboard inside the story has scroll bars:

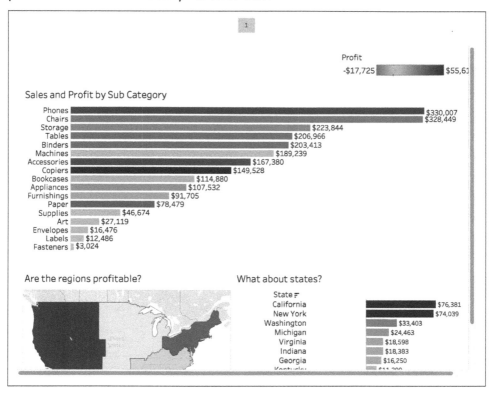

This is because we have changed the size of the story, which affects the size of the available screen size for the dashboard. To make sure the dashboard doesn't have scroll bars within the story point, we need to go back to the dashboard and change the size to fit the story:

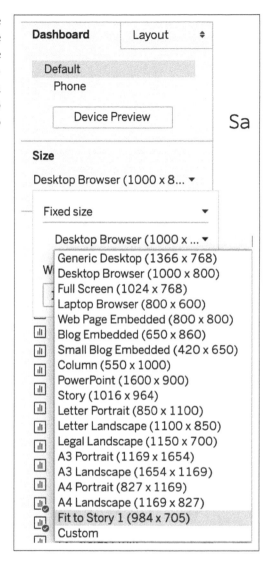

Then when you go back to your story, you will see that the dashboard fits nicely within the confinements of the story:

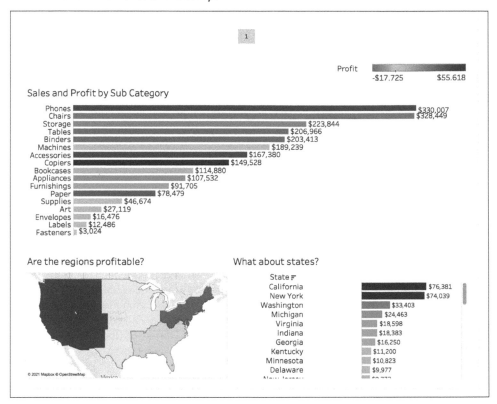

 Whenever you change any of the formatting inside a story that changes the size of the confinement area, you will have to change the size of the dashboard(s) that sit in each of the story points.

Formatting the story prior to any analysis is very important to make sure you don't have to replicate any steps during the creation. Once the story is formatted, it is now time to start your analysis. There are several ways to do this.

13.2 Adding a New Story Point

If you have several dashboards you want to use in your analysis, you will want to be able to add a new story point for each dashboard or worksheet.

Problem

You want to add a new story point with a worksheet.

Solution

In the Story pane, find the sheet you want to add to the story on the left, and double-click it. This will automatically bring this worksheet in as a new story point:

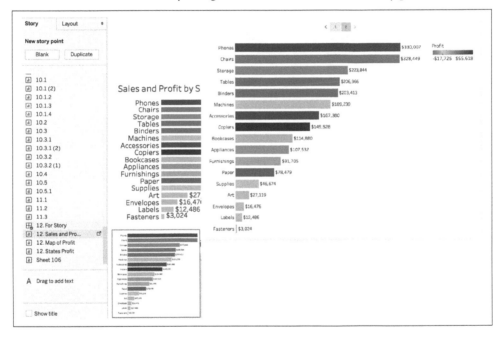

Discussion

You can add a new story point to your story in many other ways. One of the other ways is using Tableau's drag-and-drop functionality. To do this, you need to find the sheet you want to add and drag and drop it next to the first story point:

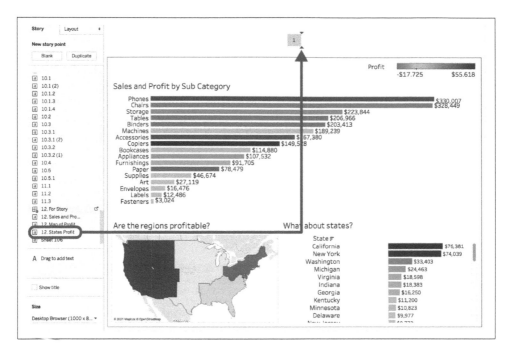

If you want to replace a story point with a different sheet or dashboard, you can drag and drop it into the middle of the story point:

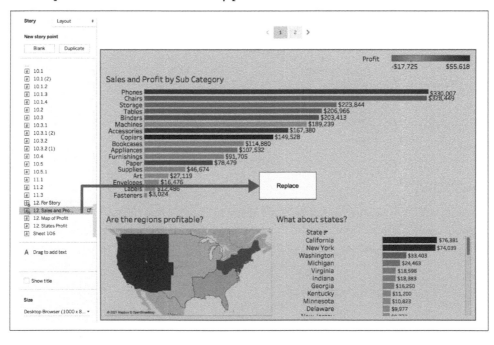

Two other options for adding a new story point are to add a blank story point and to duplicate a story.

13.3 Blank Story Point

Problem

You want to add a blank story point in order to add a starting point with just a title.

Solution

1. Under the Story tab, in "New story point," select Blank:

2. Select number 3 and move it to the front of the story points:

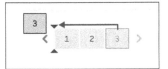

3. Add a text box to add the title, which will give you a pop-up box to edit the text:

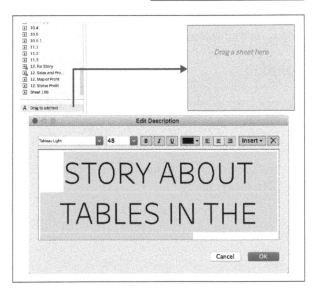

The story should now look like this with three points:

Discussion

Adding blank story points allows you to add narrative throughout the story.

13.4 Duplicate Story Point

Problem

You want to duplicate a previous story point.

Solution

1. Select which story point you want to duplicate:

2. In the Story pane on the left, select Duplicate under "New story point":

Discussion

Duplicating a story point allows you to keep any previous edits of a story point in the new one.

13.5 Annotations on a Story

The advantage of using a story instead of dashboards is that it allows you to guide a user through a specific piece of analysis. Any filters, comments, or annotations you do within a story are not reflected in the original dashboard and are specific to the story. Using annotations allows you to guide a user through a journey without having to talk about it.

Problem

You want to add an annotation to the story point to show the user where to focus.

Solution

1. Right-click a mark and select Annotate > Mark:

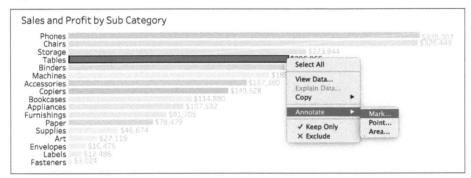

This allows you to include specific information about the mark you have selected in your annotation:

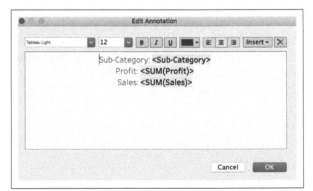

2. I recommend forming a
   sentence to highlight the
   key points of the mark
   you have selected:

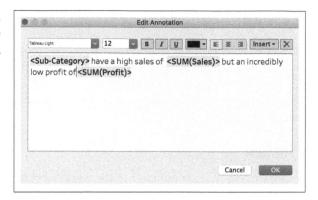

This is what your annotation will look like in the story:

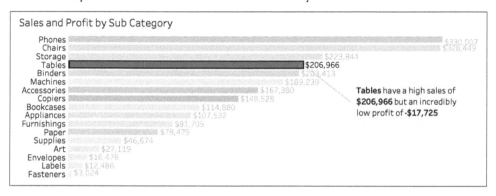

Discussion

Adding annotations is a great way to guide your user through the analysis you have
discovered. When you are creating your story and adding annotations, you might
forget to duplicate the story point before annotating to show the bigger picture before
drilling into a specific annotation. For this reason, Tableau has implemented the
option to "Save as New."

13.6 "Save as New" Story

The "Save as New" story feature allows you to duplicate the current story point with the annotations/filters, etc., while also keeping the original story.

Problem

You want to keep the original story point along with the new annotated story.

Solution

After you have annotated or made changes to the original story point, select the option above the story point: "Save as New."

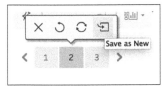

This creates a new story point with your annotations and filters:

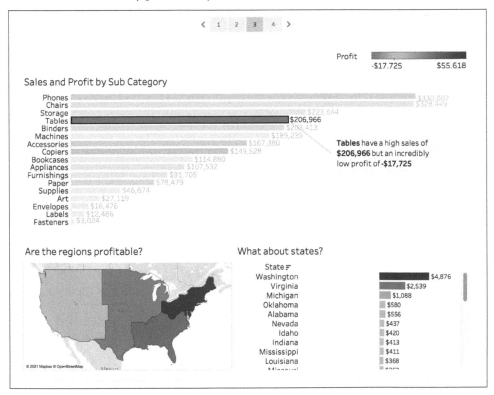

This also keeps the original story point without the annotations:

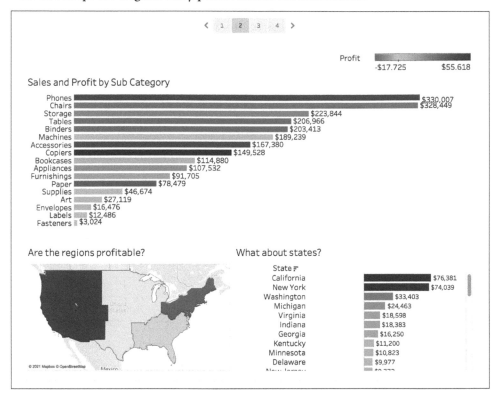

Discussion

"The Save as New" story option allows you to keep within the flow of guiding your user. Each time you make a change or a selection, you can save as a new story point that will maintain any selections you have made in that point.

The other options for changing a story point are to Revert to the original story point or to Update to reflect new changes.

A story is a great way to guide the user through a piece of analysis without being with them. It can be used to allow for deeper insights and investigation of the data during presentations.

Summary

In this chapter, you have built your first story. Stories allow you to guide a user through some analysis without being with them, by using annotations and text objects. Stories are underused and could be used to replicate a presentation, with the ability to still filter and interact with the dashboards/worksheets inside a story. You might find that the story enables your viewers to ask more questions about the analysis.

Part III Conclusion

You are three-quarters of the way through this book. In this part, you have continued to build many chart types, used different mark types, and finished off with creating a story. Part IV will enable you to use techniques to enhance your Tableau analysis and create more interactivity within your visualizations.

Advancing Your Techniques

CHAPTER 14

Sets

Previous chapters have covered the basics on how to build chart types and different functionalities. The next few chapters are going to dig further into advanced techniques and functionalities, starting with sets.

A set is defined as a collection of elements or objects. A set in Tableau is no different; they are custom-made fields generated from a subset of data based on rules or a selection. When a set is used in Tableau, it generates a field with IN or OUT results. Any elements that meet the criteria are IN the set, and then everything else is OUT of the set.

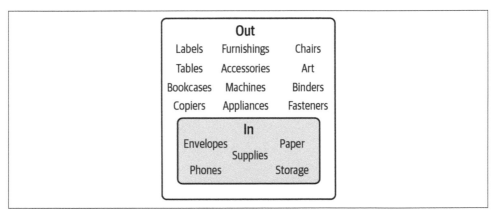

Sets are like groups, except sets can be dynamic. You can create several types of sets in Tableau, all depending on what it is you are trying to achieve with each one. Sets can be created only by using dimensions.

14.1 General Set

A *general set* uses specifically selected members within a dimension. These can be used when you always want to compare the selected fields.

Problem

You want to create a set to show a specific selection of subcategories.

Solution

1. Right-click Sub-Category and select Create > Set:

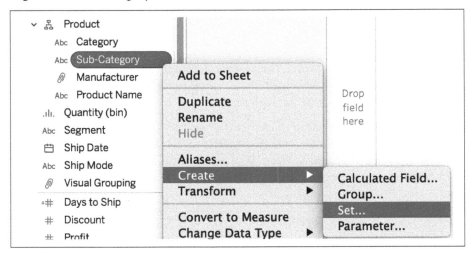

2. In the pop-up box, select subcategories that you are interested in:

3. Once you have clicked OK, you will find your set in the dimensions section on the left; this will have a Venn diagram to show that this is a set:

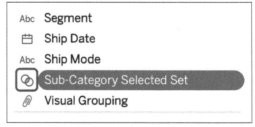

Discussion

Creating a general set is much like creating groups in Tableau. The difference is that a group allows you to name the selected group of fields and decide whether to group the rest of the nonselected fields into an "other" group or leave them as they are.

Sets, on the other hand, give you only the two options of IN or OUT of the set. If we add the newly created set to the view by double-clicking, you can see what Tableau does with this new field:

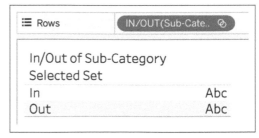

The Sub-Categories are either in or out of the set. If we add sales to the columns, we can now see how much sales the subcategories in the set had compared to those out of the set:

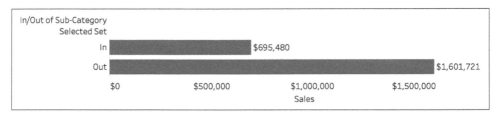

This set can now be used throughout the entire workbook. However, if you refreshed your data, any new sub-category added will automatically be placed in the OUT result.

14.2 Conditional Set

A general set will remain static if new subcategories come into the data set; however, a *conditional set* allows you to add members into the set based on certain criteria.

Problem

You want to create a set to show customers with sales greater than $10,000.

Solution

1. Right-click Customer Name and choose Create > Set:

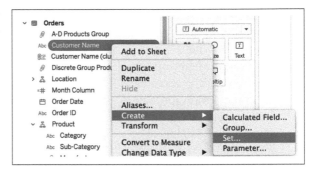

2. Rename your set. Make sure you select the "Use all" option, because we want to include all the customers before creating the condition statement:

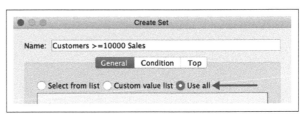

The Condition tab is where we're going to create our statement:

 The tabs within a set are not exclusive, and they work together. If you make changes to General and Condition, *both* will apply.

3. This tab has three options. The first is no condition, the second is by field, and the third is by formula. Select "By field," which has drop-down options (1). For this example, select the option for "Sales field is greater than or equal" (2) to $10,000 (3):

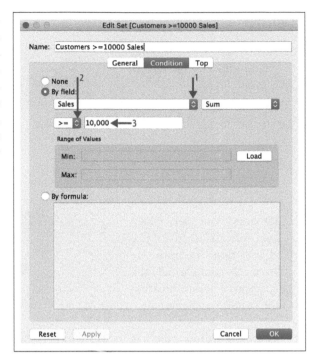

The "By formula" option allows the user to type a specific calculation criterion, which needs to result in a true or false answer.

4. Before you click OK, go back to the bottom of the General tab it to see a summary of the set and review what is being applied:

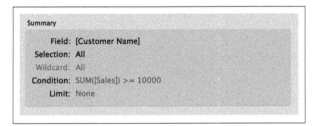

Discussion

A conditional set is dynamic. If new or existing customers have higher than $10,000 in sales, they will automatically be added to this set. You can use this set in a few ways.

You can add it to the view to create a split between in or out of the set. To do this, first add Customer Name to Rows, and Sales to Columns, and sort descending (see Recipe 3.1).

If you add the set to Color on the Marks card, you will automatically see the difference between those in and out of the set:

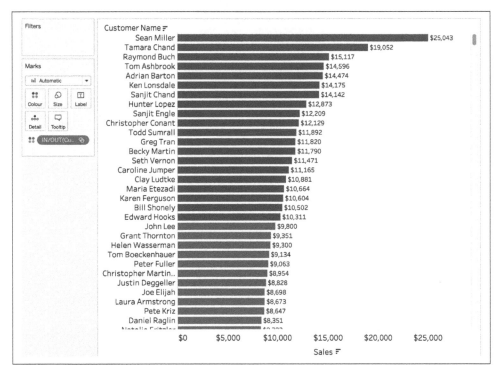

If you add the set to Filters, it will now bring back only the members in the set:

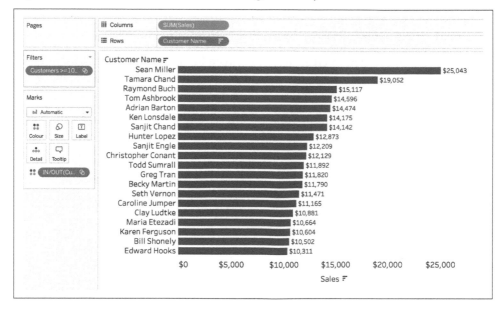

14.3 Top N Set

Another dynamic type of set uses *Top N*. Much like the conditional set, it allows you to bring back the top *n* dimensions based on a measure. For example, this type of set could bring back the top 10 customers based on the Sum of Profit.

Problem

You want to see the top 10 customers based on their profit.

Solution

1. Right-click the customer name and create a set.

2. As with the conditional set, you need to make sure you select the "Use all" option on the General tab:

3. On the Top tab, you have the same three options as the conditional set, None, "By field," or "By formula."

4. Select "By field" and use the text field and drop-downs to indicate Top 10 (1), by the field Profit (2), and Sum as the aggregation (3):

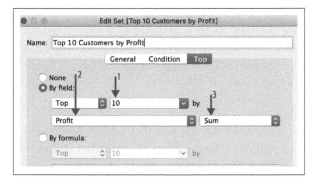

5. Go back to the General tab to see the summary of the set:

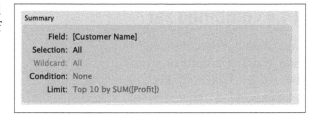

Discussion

As with all the sets, you can use this on Color on the Marks card. Add it to Filters or Rows.

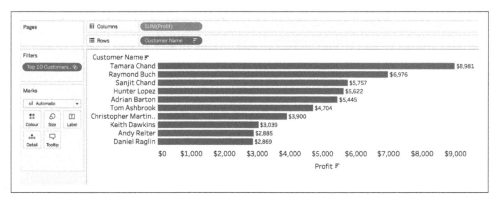

The Top N set is very similar to using the Top N filter, as mentioned in Recipe 3.3, where you can use a parameter to define the number you would like to bring back. You can do this by clicking the drop-down next to the text field:

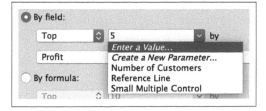

Although slightly confusing, you can use the Top section of a set to bring back the Bottom N as well. For this, you would click the drop-down next to Top and select Bottom:

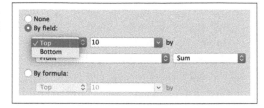

Being able to use the sets throughout different worksheets and dashboards is where the power of a set comes in, but what if you want to combine those sets to get a smaller subset of data?

14.4 Combined Set

You could use both the Condition and Top criteria in a single set by following Recipes 14.2 and 14.3 without creating a new set each time. This will show up in the Summary section to show that you have used two criteria to define this set.

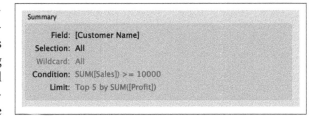

But this will mean you have several places to change if you change the criteria. To help with this, Tableau has a *combined set*, which does exactly as it says, by combining two sets together, with various connection options.

Problem

You want to create a combined set to show customers who have greater than or equal to $10,000 in sales and also fall in the top 10 for the amount of profit.

Solution

1. Press Ctrl (Command on Mac) and highlight both sets created in Recipes 14.2 and 14.3:

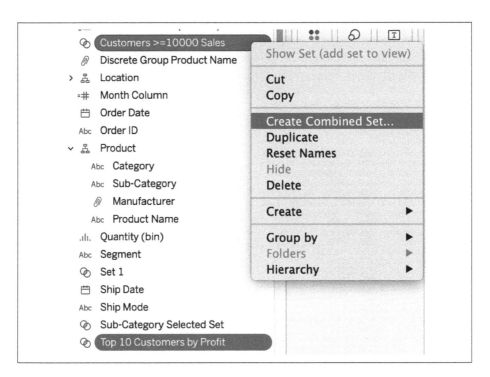

2. Give the set a sensible name and select "Shared members in both sets":

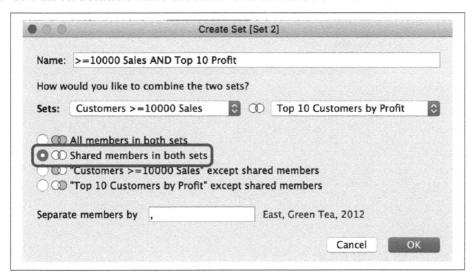

Discussion

To show you how this combined set works, you will need to create a view with Customer Name on Rows, Sales and Profit on Columns, sorted descending by profit:

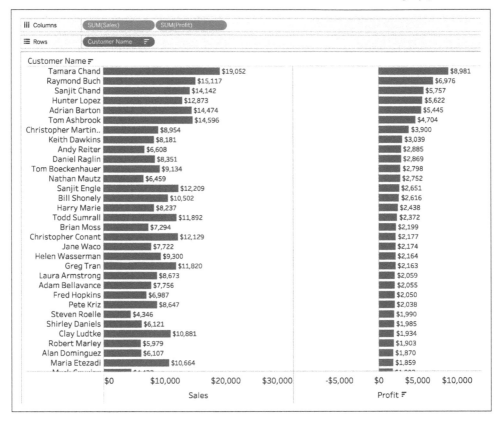

If we add the conditional set to the Sales Marks card and the top set to the Profit Marks card, you'll start to picture what to expect:

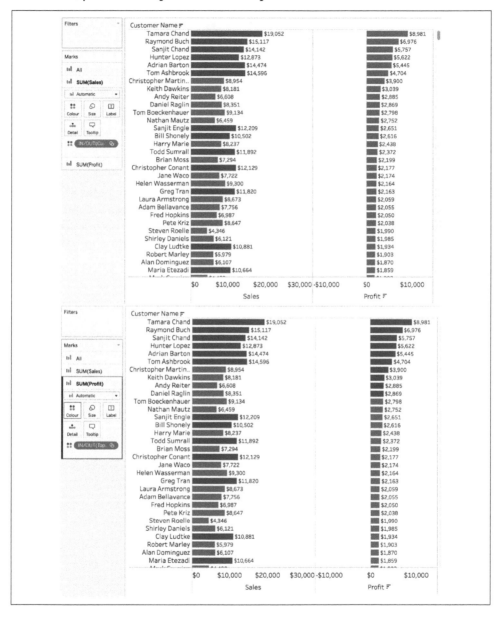

As you can see, only six of the customers fall within the top 10 by profit. If we add the combined set to Filters, we will get back only six records:

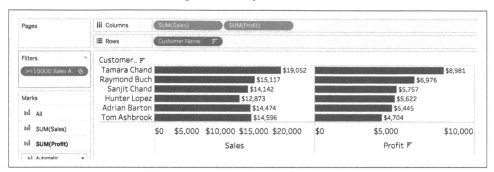

A combined set allows you to look at the members that are in both of the sets and use this within other analyses throughout the workbook. You can use different options when creating a combined set. The first option is "All members in both sets." That will bring back all the customers who have sales greater than or equal to $10,000 and those customers who are in the top 10 for profit:

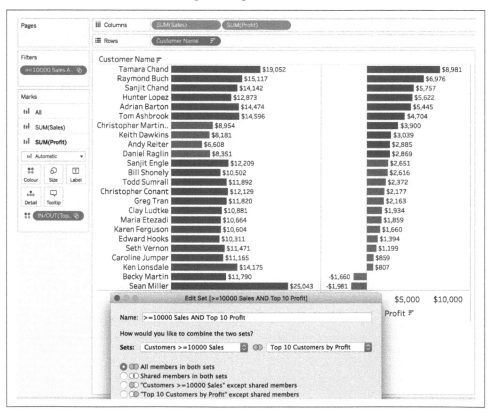

The third and fourth options refer to those members that are in only one set and do not have combined elements:

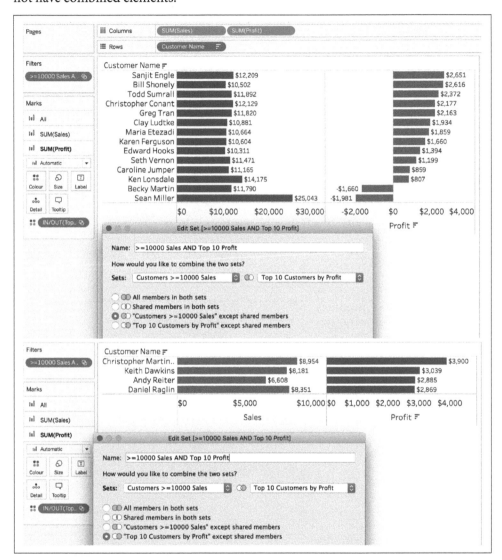

14.5 Constant Set

A *constant*, or *fixed*, set is a set you can create using one or more dimensions, but once the set is created, the only way to change its content is within a view that is using that set.

Problem

You want to create a constant set based on the position of marks on a scatter plot.

Solution

1. Create a scatter plot of sales and profit broken down by product:

2. Highlight the products higher than $0 profit and higher than $20,000 sales:

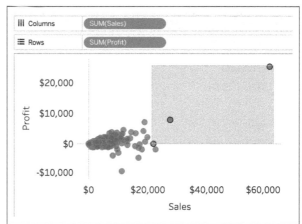

3. This will automatically bring this command pop-up. Here, you can see the Venn diagrams that represent sets; if you select the Venn diagram, you have the option to create a set:

4. Once you have selected Create Set, you will get another pop-up that looks slightly different from the other Create Set windows. This one allows only the highlighted values to be a part of the set and you can remove them only from this particular set:

5. To edit this set, you would normally right-click the set and select Edit Set to be able to include or exclude different values, but you still get only the same three values that you selected previously:

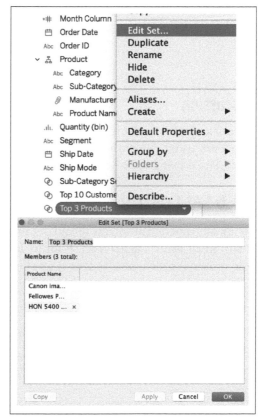

6. To add new members to this set, you will need to right-click a mark, go to the set option and select "Add to set." You will notice you have the option to also remove from the top 3 set as well:

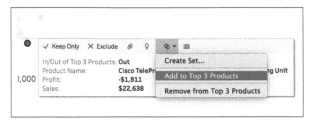

7. Finally, if you add the set to Color on the Marks card, you can see which products are in or out of this new set:

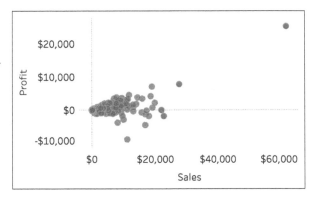

Discussion

A constant set is useful when you have static data and are using it to tell a single story. However, using one of the other types of sets allows you to have better control when your data is being refreshed, especially with the conditional and Top sets.

14.6 Set Control

Being able to control what is in and out of a set is important for some use cases, especially when using a general set. Prior to Tableau version 2020.2, the way a user could add or remove members from a set was hidden in the command menu of a tooltip:

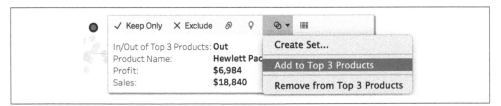

The 2020.2 version released a feature called *set controls*. These look and act like filters, except they control what is in and out of the set.

Problem

You want to give the user control over what is in and out of their set.

Solution

1. Create a bar chart with Sub-Category on Rows and Sales on Columns.

2. Add the set created in Recipe 14.1 to Color on the Marks card:

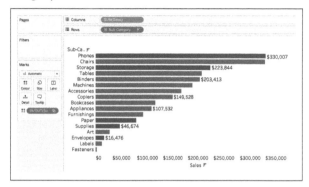

3. Right-click the set and select Show Set:

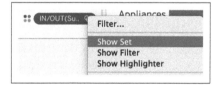

This puts a filter-like control on the right side of the sheet:

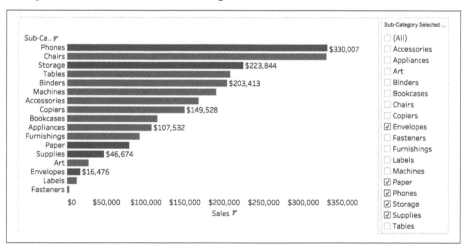

4. Now when we remove Envelopes and add Machines, you will see Tableau automatically adjusts the set and changes what is in and out of the set:

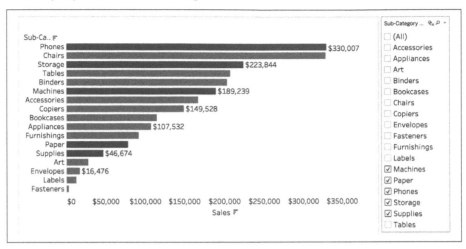

Discussion

Adding set control to the worksheet or dashboard gives the user greater control of what they want to see in the set. They might want control over a group of items, or they might be interested in particular subcategories over others.

You can customize the way the set control looks by clicking the drop-down and selecting any of the options:

Set controls add an extra card to the dashboard, but what if you want the user to be able to click a mark to add it automatically to the set?

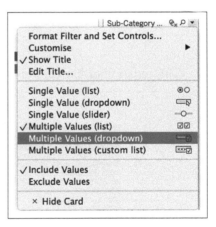

14.7 Set Actions

Set actions allow the user to interact with the view to add or remove an item from an existing set.

Top Tip 5: Set Actions

Set actions are extremely powerful in enhancing the way you can work with Tableau. Especially when comparing what is in and out of the set. Follow this recipe carefully and also see the resources for further examples.

Problem

You want to create a set action to allow the user to add a subcategory to the set.

Solution

1. Duplicate or re-create Recipe 14.6.

2. To add a set action to the worksheet, click Worksheet > Actions, which will bring up this pop-up menu:

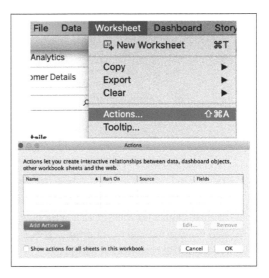

3. Click Add Action and select Change Set Values:

You will then be greeted with this pop-up box:

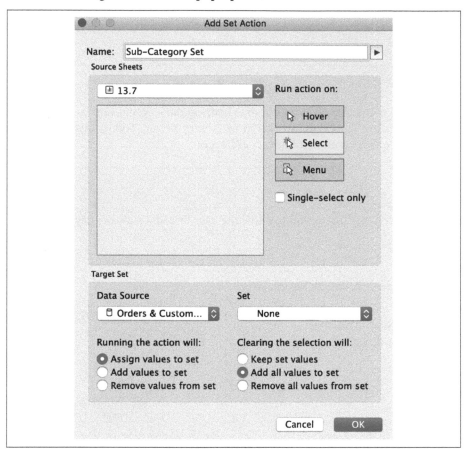

4. To create this set action, select the set (1), change the running action to add values to the set (2), which will automatically change the clearing the selection setting (3):

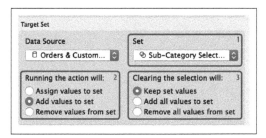

5. Finally, when you click a mark that is not already in the set, i.e., Copiers, you'll see it turns blue, meaning it has been added to the set:

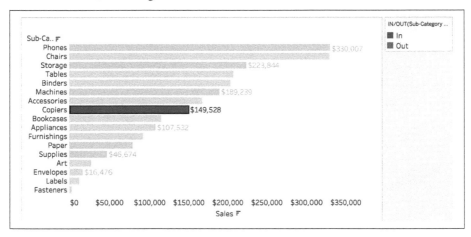

Discussion

Set actions are very powerful when interacting with charts. They can be used at the sheet and dashboard levels. The current recipe is created at the sheet level and means that if you add it to a dashboard, this action will be lost. To make sure the actions are used across the workbook, you will need to edit the action. To do this, choose Worksheet > Actions, click the ready-made action click, and Edit:

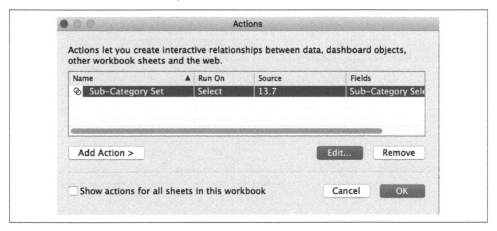

In the pop-up box, you will need to change this drop-down to be reflective of the data source:

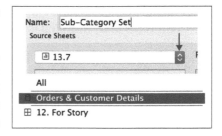

This option will bring up all the worksheets in the workbook and select them all. The action will work only on the sheets that include the set item on the Marks card. Now, once you have clicked OK and you bring the worksheet onto a dashboard, the action will still work.

 Set Actions will work only if the set is in the view, either on Rows, Columns, or on the Marks cards or reflected in a calculated field that is in the view.

One of the issues with the current setup of the action is that once you add all the sub-categories to the set, there isn't a reset button to remove them without creating a reset button on a dashboard. There are ways to avoid this, using the different options within a set action setup.

First, "Run action on" allows you to choose from three options. If you want to add members to the set when you hover over a mark, select the first option, Select is what we already used, and finally, Menu creates a hyperlink in the tooltip of the mark:

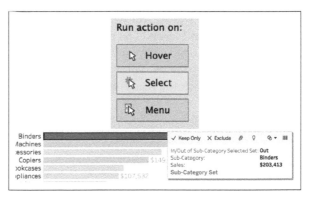

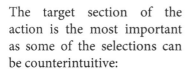

The target section of the action is the most important as some of the selections can be counterintuitive:

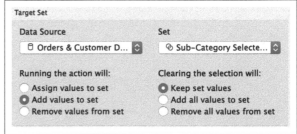

There are nine possible combinations of set actions set up. For the part of "Running the action will," you have three options:

Assign values to set
    Regardless of what is currently in the set, it will use what is selected in the view.

Add values to set
    Adds the values to what is currently in the set.

Remove values from set
    Removes any current values in the set.

On the "Clearing the selection will" option comes another three choices for what happens to the set:

Keep set values
    This will keep any set values you have in the set.

Add all values to set
    This will add all the values back into the set.

Remove all values from set
    This removes all the values from the set and returns an empty set.

All combinations are not logically possible as they will run out of values to enter or remove at a given point. Each combination would be based on a specific use case.

See Also

Ryan Sleeper, *Innovative Tableau*, Tips 95–97, "3 Innovative Ways to Use Set Actions"

Workout Wednesday 2020, Week 20: "How much do these states contribute to the total?" (*https://oreil.ly/CAjRM*)

Workout Wednesday 2020, Week 25: "Can you add and remove items from a set?" (*https://oreil.ly/szq0b*)

Workout Wednesday 2020, Week 30: "Can you create a drill down using set actions?" (*https://oreil.ly/UwrqY*)

Summary

Using sets, set controls, and set actions allows you to create deeper analysis into the dimensions you really care about. They also give your user the ability to control aspects of the visualization.

CHAPTER 15

Parameters

Parameters are user-generated values at the worksheet or dashboard level. In some cases, they allow the user to compare against a selected value without filtering the data. Parameters have many use cases. This chapter will cover some of those uses. The benefit of using a parameter is it is independent of a data source and is relevant to the whole workbook, because until you tell Tableau what to do with the parameter, it stays independent of the data set(s).

Tableau has several types of parameters; some have already been mentioned throughout this book.

15.1 Float/Integer Parameter

Problem

You want to give the user the ability to select a value to show subcategories above and below the value.

Solution

1. Click the drop-down menu and select Create Parameter:

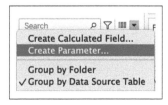

2. In the pop-up box, choose either Float, which includes decimals, or Integer, which is whole numbers:

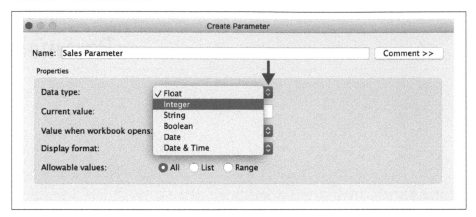

3. Add in a current value, which means what the parameter value will start on:

4. If you want to change the display format to include a dollar sign, click "Display format" and change to Currency:

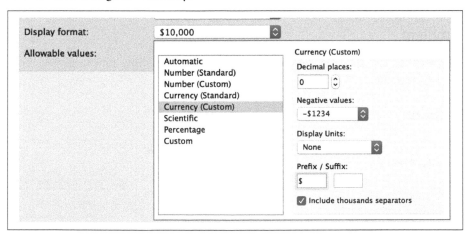

5. Then click OK to create your parameter.

The parameter will be located at the bottom of the Data pane:

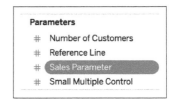

6. To show the parameter, displaying the current value, right-click and choose Show Parameter:

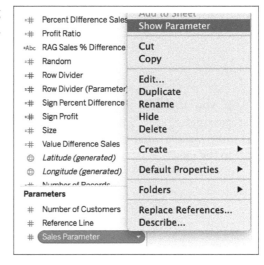

This will now appear as a type in a box on the right. If you change the value in this box, nothing will happen because we need to tell Tableau how to use this parameter.

7. Create a calculated field that looks at whether the sum of sales is greater than or equal to the value of the parameter:

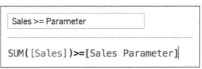

8. This calculation can now be used in any view. If you add it to Color, you can see which subcategories are above and below the sales parameter:

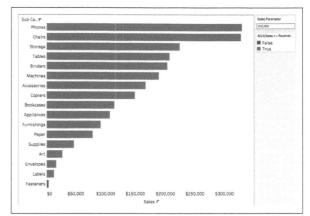

9. If you increase the sales parameter to $150,000, you can see what happens to the view:

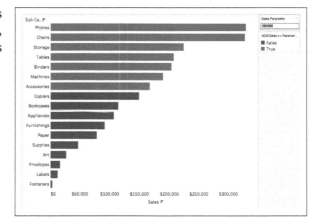

Discussion

This is one use of parameters, controlling the value reference. This could be used to compare against a user-inputted target.

There were two editable items in a parameter not mentioned in the recipe: "Values when workbook opens" and "Allowable values."

"Values when workbook opens" allows you to use a field in the data set to control the parameter whenever the workbook is first opened (more about this later when using a date parameter):

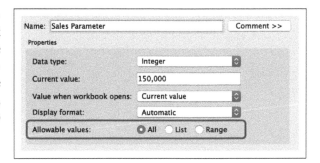

The second option is "Allowable values." This is dependent on what you want to allow the user to input into the parameter. The All option allows the user to put in any value, string, date, etc., with no limits to how big or small the value is:

The List option allows the analyst to give the user a list of options to select from, for example, allowing the user to have only a certain amount of different values:

This option also allows you to add values from a field and to update the list of values when the workbook is opened (for more about using these options, see Recipes 15.2 and 15.3).

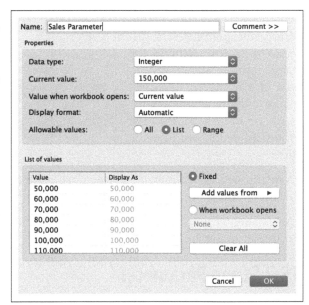

The final option is Range, which allows you to enter minimum and maximum values, but also enter a step size, so the user can scroll through a range of values based on values you have in this parameter:

Each type of allowable value and data type will allow you to display the parameter in different ways. For example, the Range option allows you to type in a number (which will accept only values within the range or default back to the minimum value), or use the slider:

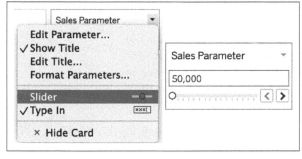

As mentioned, the difference between float and integer is that a float will allow you to include decimals, whereas an integer is whole numbers only.

See "Changing to a User-Controlled Reference Line" on page 95 for how to use a parameter as a reference line.

15.2 String Parameter

Another option for the parameter is using a string. *String parameters* allow you to select a certain string that can be used as a highlighter rather than as a filter.

Problem

You want to highlight a certain subcategory to easily see where it sits in comparison to other subcategories.

Solution

1. Right-click Sub-Category and select Create > Parameter:

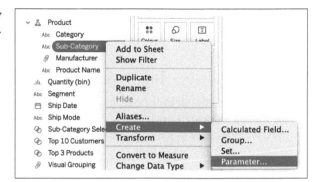

2. Creating the parameter this way allows you to autopopulate the list values with the subcategories:

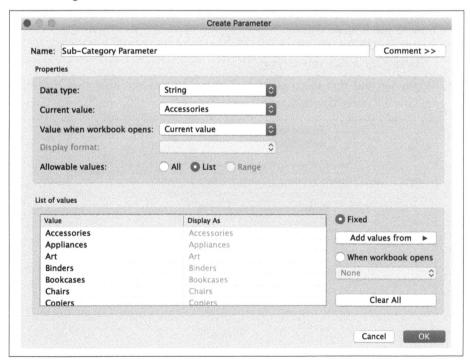

3. But if you want to create the parameter from scratch, you can still add the list by using either the Fixed or "When workbook opens" option on the right. If you know you will have a new subcategory in the data, it's best to use "When workbook opens" to allow it to be based on the current data:

Top Tip 6: Dynamic Parameters

When you use the default or fixed options, the parameters are static. But when you use "When workbook opens," it will retrieve new allowable values dynamically. Using this option is called *dynamic parameters*.

4. Once you have created your parameter, you can show the parameter, and we need to tell Tableau how to use it again with a simple calculation like "subcategory equals subcategory":

> SubCategory Highlight
>
> [Sub–Category]=[Sub–Category Parameter]

5. When we add that calculation to Color, you can see how this highlights the specifically selected subcategory of Accessories:

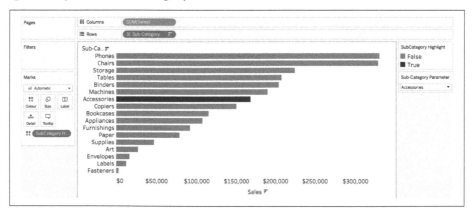

Discussion

In this example, we have used a list of values, because that way you know the user can select only what is in the data. There are still different ways you can show this parameter card because we have chosen the string data type. If you click the drop-down next to the parameter, there are different view types.

When using a string parameter, it would always be best to use a list of values:

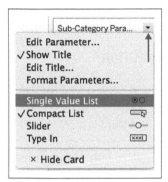

The single value list and compact list will look like a filter to any other user:

Using the list values is acceptable when you have a small number of values. If you have a massive list of options, searching through a list value is not the best option. Using the All option will allow the user to type in a specific value, which may or may not be in the list of values.

One of the downfalls of parameters is being able to select only a single value by default, which is where set, set control, and set actions would be used (Chapter 14). A parameter is more functional when you have a single dimension you want to select.

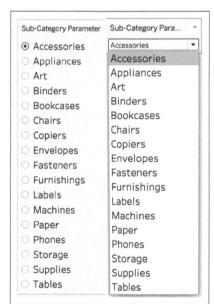

15.3 Date Parameter

Another data type that you can assign to a parameter is a *date*. This allows the user to select a date to compare against.

Problem

You want to give a user an option to select a specific month of a date.

Solution

1. Create a custom date that rolls the order date up to specific months. Right-click Order Date and choose Create > "Custom date." In this box, you want to set the Detail option to Months and keep Date Value selected:

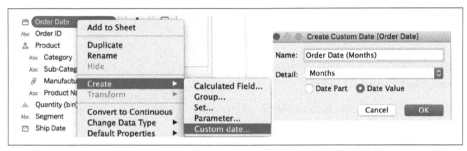

2. Right-click this new field and select Create > Parameter:

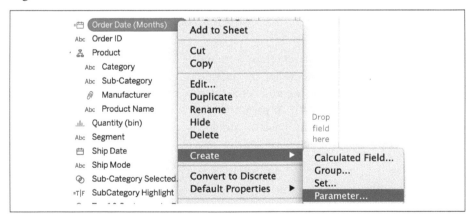

This will once again autopopulate the range values with the minimum and maximum months and years in the field:

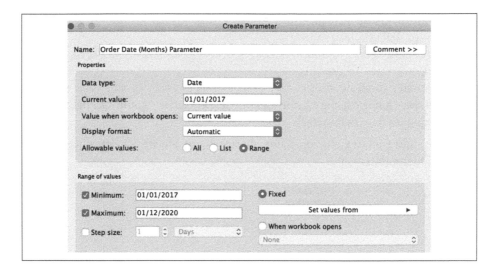

3. If you know the date field will update when new data is added, it's best to use the "When workbook opens" and select the Months field:

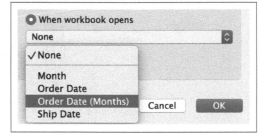

4. Click OK and show the parameter control; notice that it defaults to a slider:

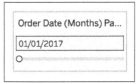

5. If you want to give the user a drop-down option to select a month and year, edit the parameter and use the List option:

6. Finally, to create a calculation to tell Tableau how this parameter should work, we want to use this parameter to show all the

sales pre- and post-parameter dates. The calculation would be "date greater than or equal to date parameter":

7. If we add this to Color on a line chart, we can see all the dates greater than the parameter date:

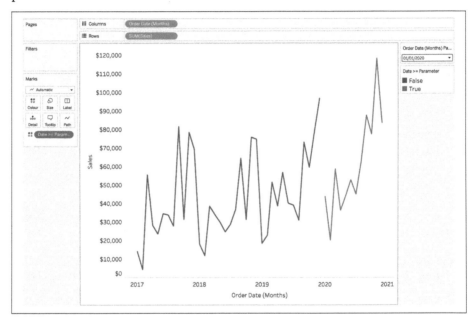

Discussion

The current value can be selected as part of the parameter, but it can also use a calculation to ensure the current value is the maximum date in the data set. To do this, we first need to create the calculation, which is a simple level-of-detail (LOD) calculation.

The curly brackets denote an LOD calculation (see Chapter 17 for more information). This is going to return the maximum month of order date in our data set:

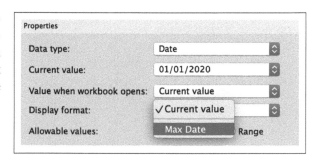

Next, we can edit our parameter, and in the "When workbook opens" drop-down, you should now be able to select the Max Date calculation we just created.

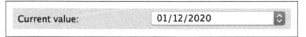

This then updates the current value to reflect what value is being brought back by the maximum date calculation.

Now when your workbook opens, it will always default to the maximum date in the data set.

Display format was mentioned in Recipe 15.1, but with dates we have different format options depending on how you want to display the date (see more about date formats in Chapter 5). When using months, I prefer to use the month name and year:

The current example gives you the option to show as a compact list, single value list, or a type-in date; however, if you change "Allowable values" to All, you can get a calendar date picker instead of the drop-down or single value list. This is better for when you have exact dates rather than at a monthly level:

15.4 Measure Switcher

There are many use cases of parameters to allow the user to explore the analysis, but there are also use cases for when you want to give your user an option to explore the data a little more. For example, if you want to change a measure in a chart, you can use a parameter.

Problem

You want to switch among several measures: sales, profit, and quantity.

Solution

1. Create a new string parameter with the following options.

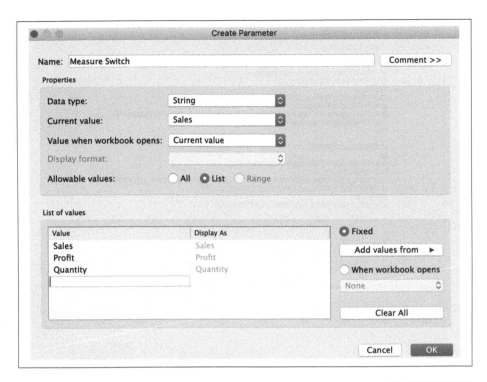

2. We now need to tell Tableau how to use this specific parameter. Create a calculation that uses this case statement:

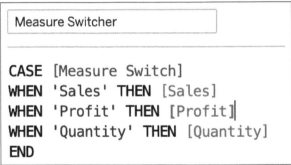

Measure Switcher

```
CASE [Measure Switch]
WHEN 'Sales' THEN [Sales]
WHEN 'Profit' THEN [Profit]
WHEN 'Quantity' THEN [Quantity]
END
```

3. Add the calculation to Columns and add Sub-Category to Rows; then turn on the Label Marks card:

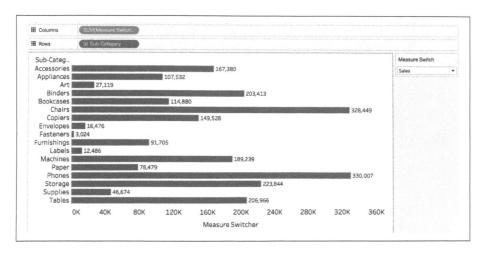

4. For this example, switch the parameter to Single Value List instead of Compact List. You can do this by clicking the drop-down arrow on the parameter control:

5. Now when we change the parameter to a different value, it will change the view to reflect that measure:

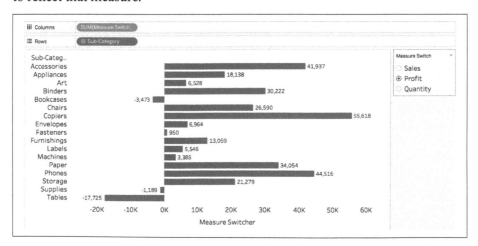

Discussion

Using the parameter in this way allows your user to interact with the view and select which measure they want to look at. It is also an excellent technique to save space on your dashboard without having to show all three measures at once.

While you have created this calculation, you can also use it as a sort by option, which will then sort the dimension descending by the highest value of that selection. Because the measure switcher is already in the view and has an axis, you can use this to sort:

15.5 Dynamic Scatter Plot

Scatter plots use two measures to compare against, but what if you want to give the user the ability to see a scatter plot of different measures on the *x* and *y* axes broken down by the products?

Problem

Create a dynamic scatter plot with interchangeable x and y axes broken down by product.

Solution

1. Follow Recipe 15.4 to get the base calculations.

2. Duplicate the measure switcher parameter and calculated field twice and label them **X** and **Y**:

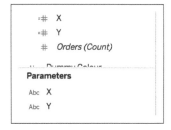

3. Edit each calculated field to reference the correct parameter:

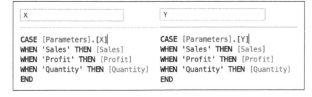

4. Add the X calculated field to Columns, Y to Rows, and Product Name to Detail:

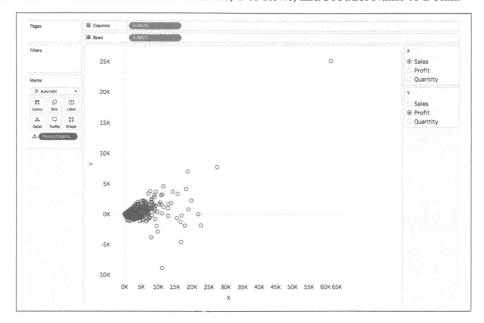

Now your user can switch the measures on each axis:

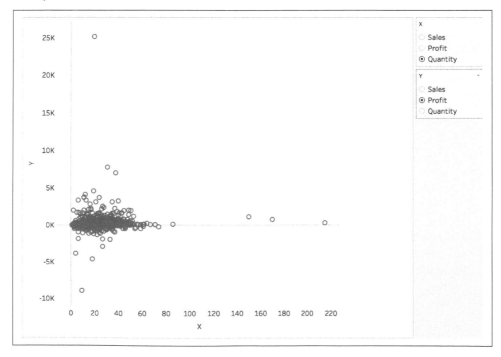

Discussion

Giving your user the ability to switch measures and explore the data is great if you are creating an exploratory dashboard.

15.6 Dimension Switcher

To take the dynamic scatter plot one step further and add another layer to the exploratory nature of this chart, you might want to give the user the option to add a dimension switch between product and customer.

Problem

You want to give the user the ability to switch dimensions between product name and customer name.

Solution

1. Create a new parameter to give the user the two options:

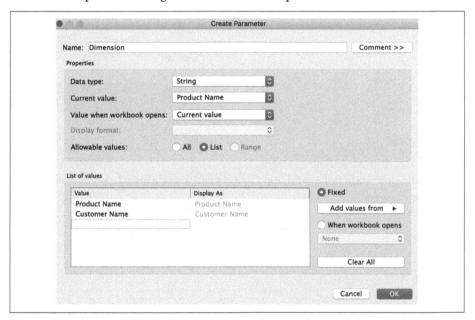

2. Create a calculation to reference the parameter:

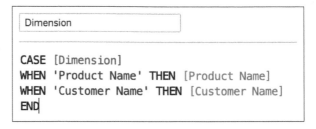

3. Duplicate Recipe 15.5 or re-create it as your baseline chart.

```
Dimension

CASE [Dimension]
WHEN 'Product Name' THEN [Product Name]
WHEN 'Customer Name' THEN [Customer Name]
END
```

4. Add Dimension to Detail:

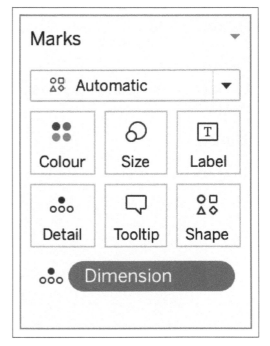

5. Show your parameter and switch between the different dimensions; watch the view change:

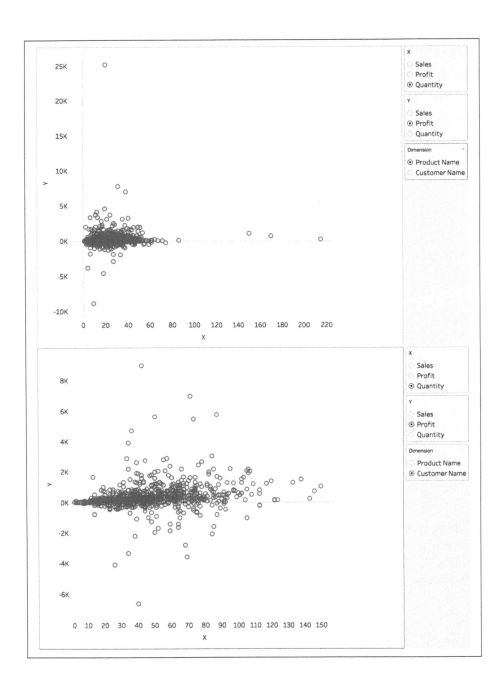

Discussion

The ability to allow your users to explore the data by using this type of chart and use case is going to drive further analytics.

To enhance the current view, I recommend adding X and Y headers by adding the relevant parameters to Columns and Rows:

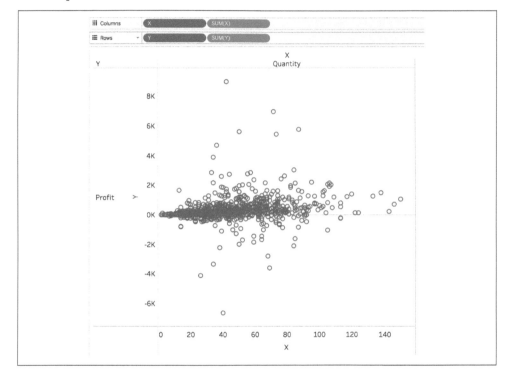

And finally, add a worksheet title to tell your user exactly what is on the X and Y, but also at what level of detail the view is. You can add parameters to titles for additional context. To do this, double-click your title, click Insert, and find the parameter Parameters.X:

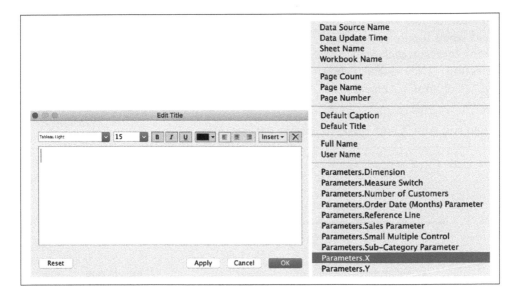

Add the relevant parameters into the title and form a sentence:

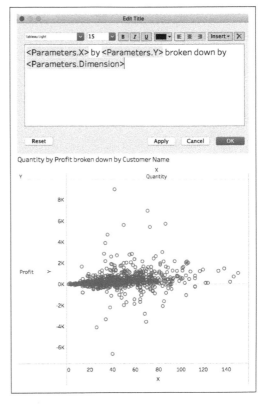

To clean up this view, hide the field labels for Rows and Columns by right-clicking X and selecting "Hide Field Labels for Columns"; repeat for Rows:

Then edit your X and Y axes to remove the axis title:

Edit Axis [Y]   ✕

General   Tick Marks

Range

⦿ Automatic   ☑ Include zero
◯ Uniform axis range for all rows or columns
◯ Independent axis ranges for each row or column
◯ Fixed

| Automatic ▾ | Automatic ▾ |

-7,554.77934879   9,909.71374879

Scale

☐ Reversed
☐ Logarithmic

⦿ Positive   ◯ Symmetric

Axis Titles

Title

[Title   ]

Subtitle

Subtitle _____   ☑ Automatic

↺ **Reset**

And then you are left with a clean view like this:

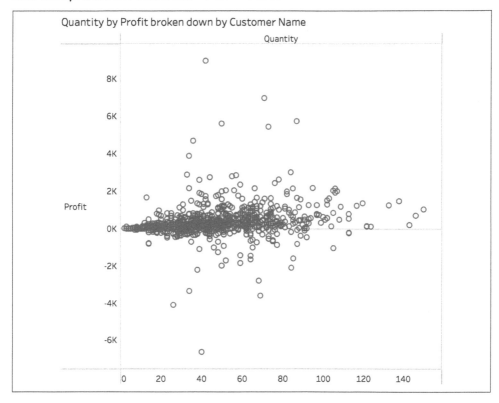

15.7 Parameter Actions

Parameters allow you to add additional interactivity to your analysis, allowing the user to select a specific reference point or mark. But what if you could use the marks in the view to change the parameter value? This is what parameter actions allow you to do. Parameter actions are very similar to set actions, except that they usually accept only a single value in the parameter. This section will show you how to create a parameter action and some practical use cases for parameter actions.

Top Tip 7: Parameter Actions

When Parameter actions were introduced they added that next level of interactivity to Tableau. The next set of recipes go through some example use cases.

Problem

You want to allow the user to select a mark to move the position of a reference line.

Solution

1. Continuing with the dynamic scatter plot, add a parameter-based reference line to each axis:

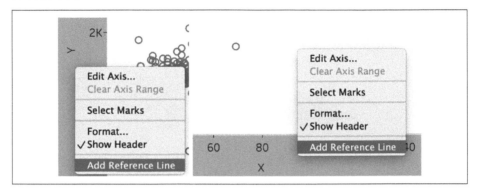

2. Select "Create a New Parameter" for each axis::

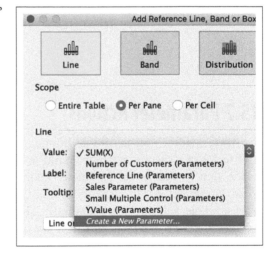

3. Name the two parameters appropriately—**XValue** and **YValue**—to allow other people to understand what each parameter is for:

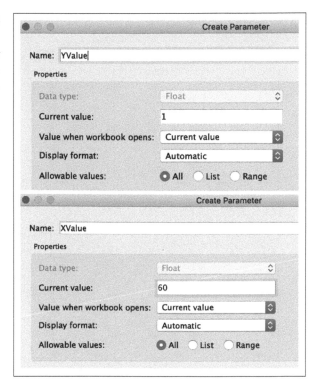

Your current view should now look like this:

4. We now want to be able to click any mark and to update both X- and Y-value parameters with the relevant values:

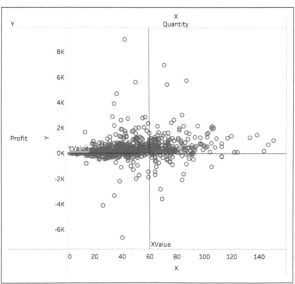

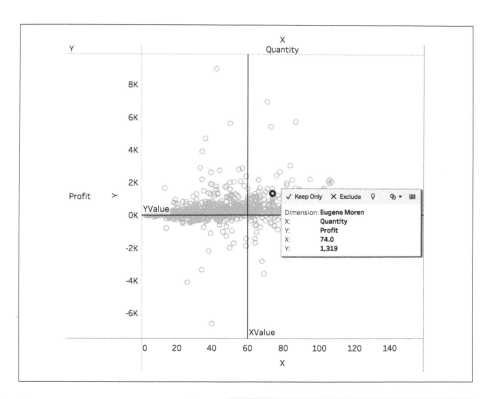

5. To create a parameter action, choose Worksheet > Actions:

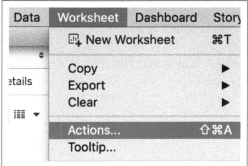

6. From here, we need to add a new parameter action for the X value and Y value. Select Add Action > Change Parameter:

This pop-up box will then appear:

7. As with the set actions, we need to tell Tableau which parameter we want to affect:

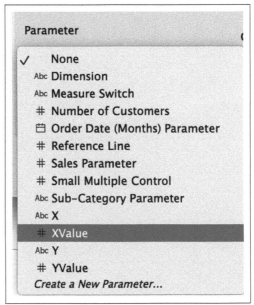

8. We then need to select which field to use to affect that parameter:

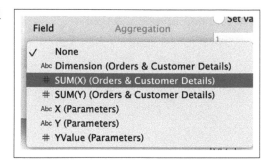

This is what your final settings should look like for the X-value parameter:

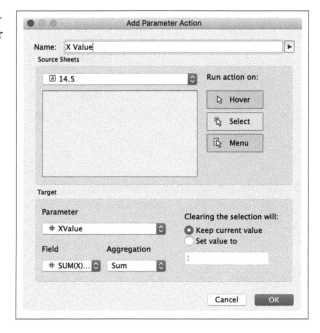

9. Repeat steps 6–8 for the Y value:

10. Click OK and OK again to get you back to the chart view.

11. To show how this action is working, select both Parameters and Show Parameter:

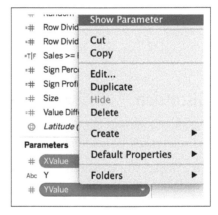

12. Notice that the values are set at 1 and 60:

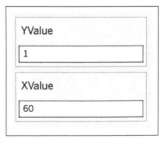

13. Now when we select any mark in the view, it will take the X and Y value and change the parameter to match, which will subsequently update the reference lines that are based off of the parameter:

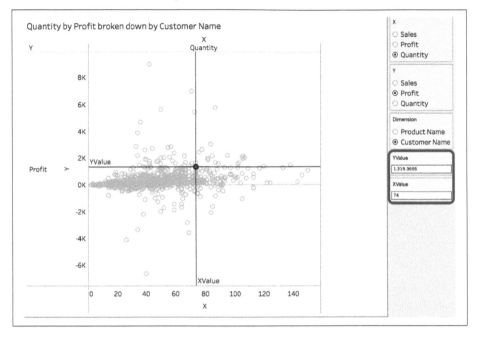

Discussion

Parameter actions allow the user to use the marks in the view to change the parameter references. This adds more interaction to your views.

To enhance this dynamic scatter plot further, I would create a calculation to show the quadrants by color, indicating which customers are above both the X and Y parameters. To do this, you will need to create the following calculation:

```
Quadrant

IF SUM([X])>=[XValue] AND SUM([Y])>= [YValue] THEN "High High"
ELSEIF SUM([X])<=[XValue] AND SUM([Y])>= [YValue] THEN "Low High"
ELSEIF SUM([X])>= [XValue] AND SUM([Y])<= [YValue] THEN "High Low"
ELSEIF SUM([X])<= [XValue] AND SUM([Y])<=[YValue] THEN "Low Low"
END
```

This calculation looks at all the possible combinations and gives it a category. When we add this calculation to Color, it makes the quadrants easier to see:

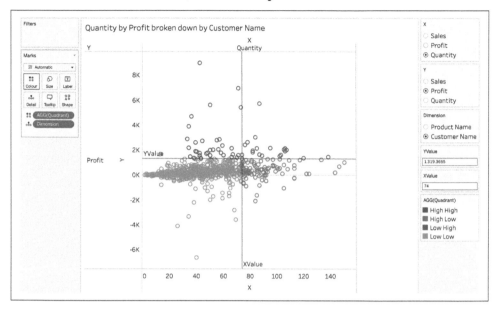

See Also

Ryan Sleeper, *Innovative Tableau*, Tips 69–71, "3 Innovative Ways to Use Parameter Actions"

Workout Wednesday 2020, Week 7: "What happens if? Can you update sales forecasts and targets using only parameters?" (*https://oreil.ly/A1uln*)

Workout Wednesday 2019, Week 36: "Can you build a custom axis with a tracking reference line?" (*https://oreil.ly/oES2h*)

15.8 Parameter Actions: Drill Down

You can use parameter actions in many ways to enhance the interactivity of your views.

Problem

You want to give the user the ability to click a category to drill down to the Sub-Category.

Solution

1. Create a bar chart that is Category by Sales:

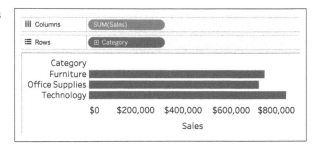

2. When a user selects Furniture, you want them to see the Sub-Categories in Furniture.

The first step is to create a category parameter. Right-click Category and select Create > Parameter:

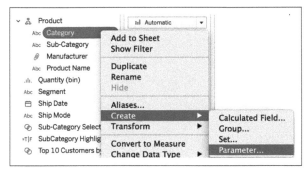

3. This gives you the categories already filled out at the bottom, but for this to work, we would want to use the All option, not List:

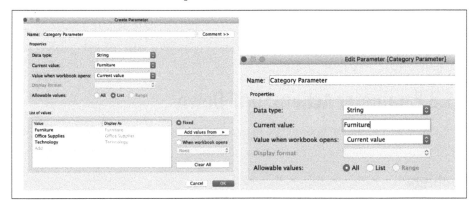

4. Don't forget to choose Show Parameter to see what Tableau is doing when we create the parameter action:

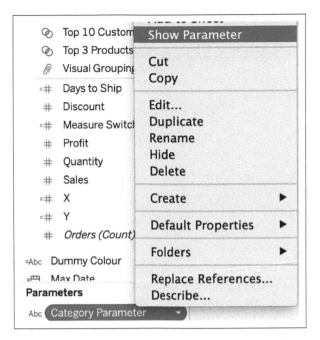

5. Before creating the parameter action, let's create the calculation needed to perform the drill-down. We want to say, "If the Category Parameter equals the Category, then give me back the Sub-Category else blank":

SubCat Drill Down

```
IF [Category]=[Category Parameter]
THEN [Sub-Category]
ELSE '' END
```

6. Then add this to the view, next to Category:

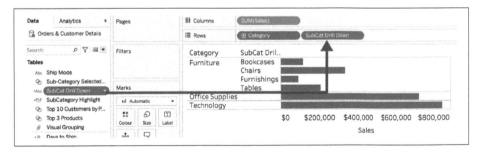

The Furniture category is already drilled down because that is what the parameter currently says.

7. To create the parameter action, choose Worksheet > Actions > Add Action > Change Parameter:

8. We need to tell Tableau which parameter and field to use in this action. We want it to affect the Category Parameter using the Category field:

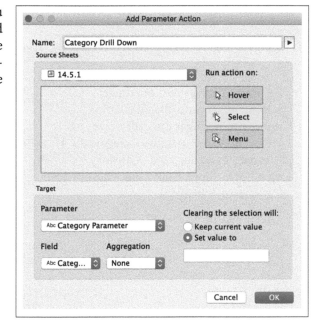

9. However, this time, you will want to change "Clearing the selection will" to "Set value to." This can stay blank. When you click off of a category, Tableau will clear the parameter and remove the drill down:

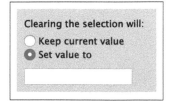

10. Once we have the parameter action set up, if you select a different Category like Technology, Tableau should update the parameter and change the drill-down view:

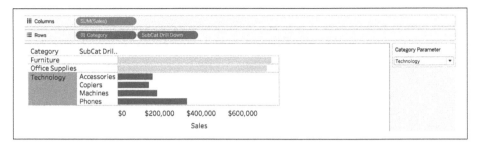

11. Now, when you click off of the category, Tableau will make the parameter blank and remove the drill-down:

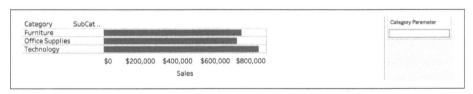

12. Finally, hiding the field labels for rows will make this view cleaner:

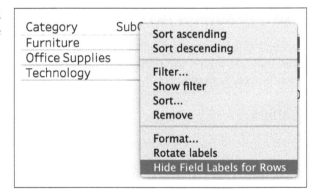

Your final view should look like this, with and without the drill-down:

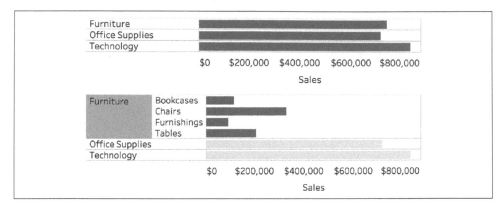

Discussion

By using the parameter actions, drill-downs are easier to build, manage, and maintain, because you are reliant on only the data that is in the view. The benefit of using parameter actions is to give the user the ability to drill down to certain subcategories, especially if short on space on the dashboard.

15.9 Parameter Actions: Min and Max Date Range

As mentioned, parameter actions allow the user to click marks to change a parameter. This use case allows the user to highlight a range of marks to give a minimum and maximum date range.

Problem

You want to allow the user to select a date range by using parameter actions.

Solution

1. Create two date parameters, a start date and an end date:

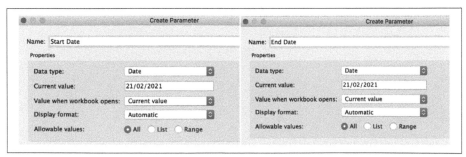

2. Always show your parameter controls before creating calculations to use them:

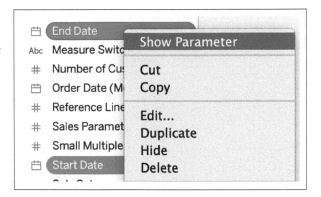

3. Create a continuous month line chart by sales:

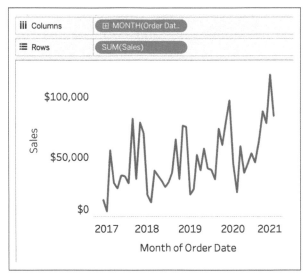

4. Create a reference band using the start and end date parameter:

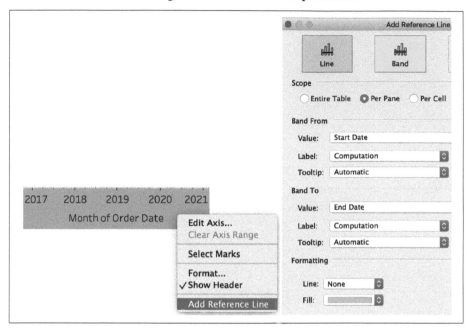

The reference band doesn't look like much yet because of the parameter being set at the same date value:

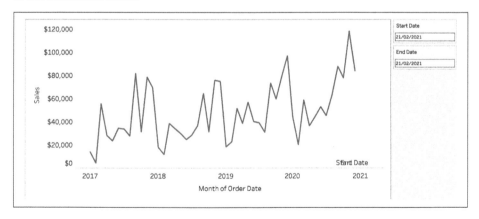

This is where parameter actions come into play. What we want to achieve is that if you highlight a date range on the view, it will update the parameter values:

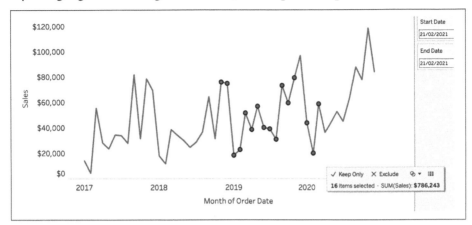

5. We need to create two parameter actions to reflect this range. Go to Worksheet and choose Actions > Add Action > Change Parameter:

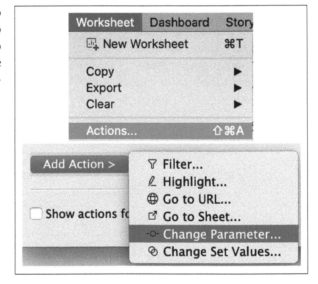

6. In this configuration, we need to first give it a name (1). Then choose which parameter we want to change (2) and with which field (3). Finally, change the aggregation to the minimum value (4):

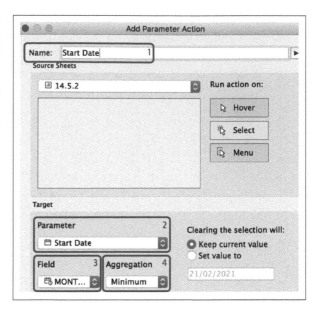

7. Repeat step 8, but using the End Date Parameter and Maximum aggregation to get the highest date selected:

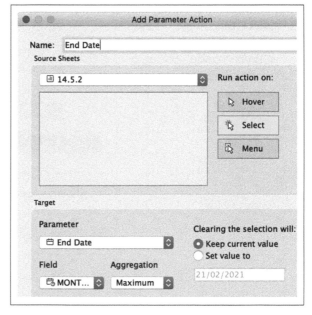

8. Now we have both parameter actions in place. When we highlight the marks in the view, you will see the reference band update to reflect those values:

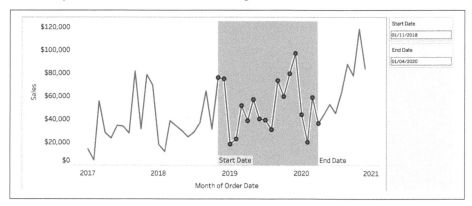

Discussion

Using parameter actions in this way gives the user the ability to select their own date range that can be used in further analysis throughout the workbook. These parameters can now be used in calculations to bring back only data within that date range.

Summary

Parameters allow your end users to really explore the data and, with parameter actions, interact with the individual marks. Parameters and parameter actions have many use cases.

CHAPTER 16

Advanced Table Calculations

Throughout this book, I have mentioned various types of Quick Table Calculations, which are available as an option to add to any continuous data field. A plethora of Table Calculations can be used in a calculated field.

Any Quick Table Calculation can be saved, which allows you to see the underlying Table Calculation involved. For example, look back at Recipe 3.5, on the Sum of Sales:

We can drag and drop the field to the Data pane on the left to save the calculation:

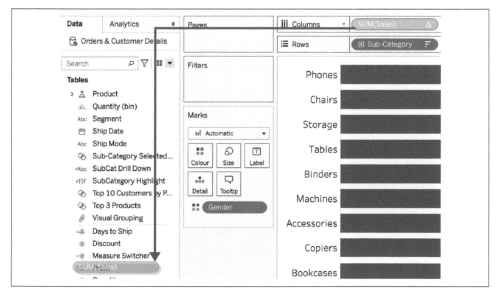

This will give you the option to name the calculation:

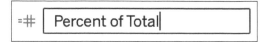

Then, once you have saved the calculation, you can right-click and edit the calculation to see what is happening under the hood:

Percent of Total

Totals summarise values from Gender.
SUM([Sales]) / TOTAL(SUM([Sales]))

In this chapter, we are going to cover some of the advanced Table Calculations and write them from the calculation window, instead of using Quick Table Calculations.

16.1 Previous Value

When creating analysis in Tableau, one use case is being able to compare to a previous value within your data. This can be seen as a difference from Quick Table Calculation, but being able to understand how to write this calculation by hand helps you fully understand how Tableau works.

Problem

You want to bring back the previous sales value when comparing to the first value in the view, so you can work out whether the current value is above or below the first.

Solution

1. Create a table with Month Year and Sales:

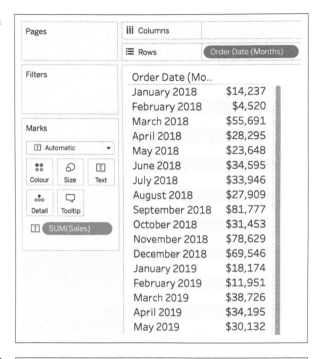

2. Open a new calculated field window and in the search pane on the right, click the drop-down to show Table Calculation, then select PREVIOUS_VALUE:

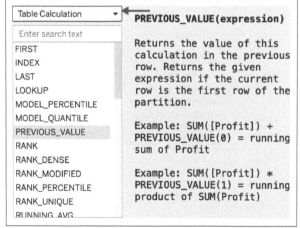

3. This gives you two examples of how the PREVIOUS_VALUE calculation works. Double-click PREVIOUS_VALUE to add it to your calculation window:

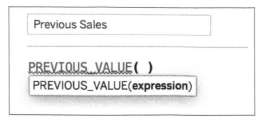

4. We have the red squiggle because we haven't told Tableau what calculation to look up for the previous value. In between the brackets, use the expression SUM(Sales):

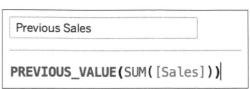

5. Once we click OK, drag the calculation on top of the sales in the view:

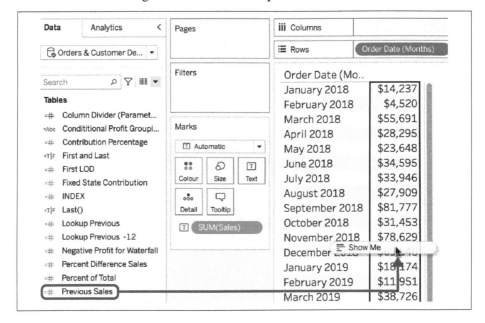

This gives you the Measure Names and Measure Values table:

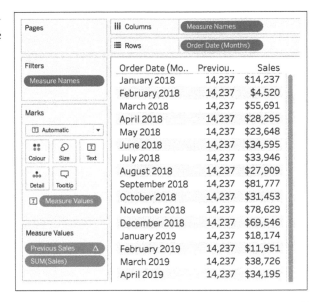

6. Notice that this value gives you the first value of January 2018 repeated across every month date. You can now create calculations that calculate the difference:

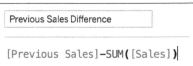

7. Double-click the new calculation to add it to the Measure Values shelf, and it shows how this calculation is working:

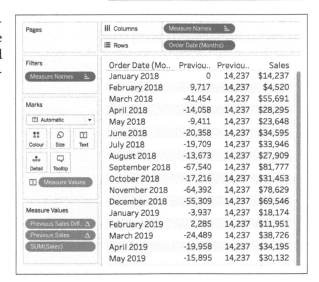

Discussion

This Table Calculation is used when you want to compare to the first value in the view. If you want to compare to the first value in each year, you can add Year to the view, and change the Compute Using of the calculation to "Pane (down)":

Pane (down) means the calculation will restart every pane, which in this case is every year. Hence, this is why the first month of every year is 0 for the difference.

				Rows		
				⊞ YEAR(Order Date)	Order Date (Months)	

Year of ..	Order Date (Mo..	Previou..	Previou..	Sales
2018	January 2018	0	14,237	$14,237
	February 2018	9,717	14,237	$4,520
	March 2018	-41,454	14,237	$55,691
	April 2018	-14,058	14,237	$28,295
	May 2018	-9,411	14,237	$23,648
	June 2018	-20,358	14,237	$34,595
	July 2018	-19,709	14,237	$33,946
	August 2018	-13,673	14,237	$27,909
	September 2018	-67,540	14,237	$81,777
	October 2018	-17,216	14,237	$31,453
	November 2018	-64,392	14,237	$78,629
	December 2018	-55,309	14,237	$69,546
2019	January 2019	0	18,174	$18,174
	February 2019	6,223	18,174	$11,951
	March 2019	-20,552	18,174	$38,726
	April 2019	-16,021	18,174	$34,195
	May 2019	-11,958	18,174	$30,132
	June 2019	-6,623	18,174	$24,797
	July 2019	-10,591	18,174	$28,765
	August 2019	-18,724	18,174	$36,898
	September 2019	-46,422	18,174	$64,596
	October 2019	-13,231	18,174	$31,405
	November 2019	-57,798	18,174	$75,973
	December 2019	-56,745	18,174	$74,920

You can also use Previous Value in a different way. Write the following syntax:

```
PREVIOUS_VALUE(0) + SUM([Sales])
```

This calculation uses the previous value and adds the current value's sum of sales. When we add this to our table, you can see what the calculation is doing:

Order Date (Months)	Sales	Previous Value along Table (Down)
January 2018	$14,237	14,237
February 2018	$4,520	18,757
March 2018	$55,691	74,448
April 2018	$28,295	102,743
May 2018	$23,648	126,391
June 2018	$34,595	160,987
July 2018	$33,946	194,933
August 2018	$27,909	222,842
September 2018	$81,777	304,620
October 2018	$31,453	336,073
November 2018	$78,629	414,702
December 2018	$69,546	484,247
January 2019	$18,174	502,422
February 2019	$11,951	514,373
March 2019	$38,726	553,099
April 2019	$34,195	587,294
May 2019	$30,132	617,426

But what if you want to calculate it to the previous month or the previous month of the previous year?

16.2 Lookup

If you want to compare to the previous month or the previous month of the previous year, you will need to use a different Table Calculation, which is Lookup. This calculation does exactly what it says —you can specify which row of data you want to return. Use the Calculation Search pane to find out more information about the calculation:

Previous Value is always going to look at the previous value, whereas, in Lookup you can specify which value you want to use.

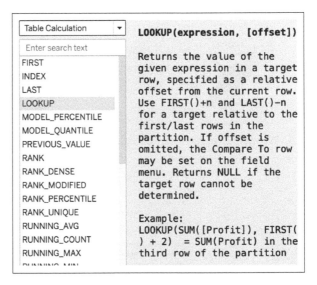

Problem

You want to bring back the previous month's sales value to compare to month-on-month values, so you can work out whether the current value is above or below the previous.

Solution

1. Start with a simple table showing Month and Year of Order Date and Sales.

2. Open a new calculation window and type **LOOKUP**:

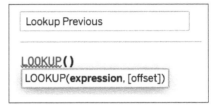

3. Once you have typed the initial calculation, you will see the syntax required for this calculation. For this example, you need an expression and an offset. Here, we will be using the SUM(Sales) with an offset of –1:

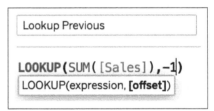

 We are choosing –1 as the offset because we want to go back one month and year to return the prior month and year value.

4. Now that we have our calculation, click OK and then add it to the view:

This is how the calculation works:

Order Date (Months)	Lookup Previous along Table ..	Sales
January 2018		$14,237
February 2018	14,237	$4,520
March 2018	4,520	$55,691
April 2018	55,691	$28,295
May 2018	28,295	$23,648
June 2018	23,648	$34,595
July 2018	34,595	$33,946
August 2018	33,946	$27,909
September 2018	27,909	$81,777
October 2018	81,777	$31,453
November 2018	31,453	$78,629
December 2018	78,629	$69,546
January 2019	69,546	$18,174
February 2019	18,174	$11,951
March 2019	11,951	$38,726
April 2019	38,726	$34,195
May 2019	34,195	$30,132
June 2019	30,132	$24,797
July 2019	24,797	$28,765

Discussion

In this example, we look up the previous value by using the offset of −1. This offset could be used to look up the next value by using an offset of 1. The offset indicates how many rows above or below the current row you want to go:

Order Date (Months)	Lookup Previous along Table ..	Sales	Lookup Post along Table (Down)
January 2018		$14,237	4,520
February 2018	14,237 -1	$4,520 +1	55,691
March 2018	4,520	$55,691	28,295
April 2018	55,691	$28,295	23,648
May 2018	28,295	$23,648	34,595

You can also use it to look up the previous month and year. For example, to compare January 2018 to January 2019, change the offset to −12 by duplicating the calculation called Lookup Previous and editing the offset to −12:

Lookup Previous -12

Results are computed along Table (across).
LOOKUP(SUM([Sales]),−12)

Now when you double-click the calculation and add it to the view, you can see how this new offset calculates the value:

When using the offset −12, you'll notice that the first 12 rows have no data. This is because Table Calculations consider only the data in the view, and there are no rows prior to January 2018 to be able to calculate this example through December 2018.

Order Date (Months)	Lookup Previous along Table ..	Sales	Lookup Previous -12 along Table ..
January 2018		$14,237	
February 2018	14,237	$4,520	
March 2018	4,520	$55,691	
April 2018	55,691	$28,295	
May 2018	28,295	$23,648	
June 2018	23,648	$34,595	
July 2018	34,595	$33,946	
August 2018	33,946	$27,909	
September 2018	27,909	$81,777	
October 2018	81,777	$31,453	
November 2018	31,453	$78,629	
December 2018	78,629	$69,546	
January 2019	69,546	$18,174	14,237
February 2019	18,174	$11,951	4,520
March 2019	11,951	$38,726	55,691
April 2019	38,726	$34,195	28,295
May 2019	34,195	$30,132	23,648
June 2019	30,132	$24,797	34,595
July 2019	24,797	$28,765	33,946
August 2019	28,765	$36,898	27,909
September 2019	36,898	$64,596	81,777
October 2019	64,596	$31,405	31,453
November 2019	31,405	$75,973	78,629
December 2019	75,973	$74,920	69,546

16.3 Window Calculations

The majority of the Table Calculations you will need are available within the Quick Table Calculation option, including the Moving Average calculation, which is discussed in Recipe 5.9.

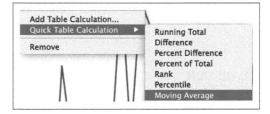

Once we have saved the calculation, we can see which functions Tableau is using to perform the calculation:

Window Average

Results are computed along Table (across).
`WINDOW_AVG(SUM([Sales]), -2, 0)`

This introduces a different type of Table Calculation known as a Window Calculation. The calculations are defined by the boundaries you set in the calculation, in this case −2, 0. This will give the average of the current and two prior values. Window calculations have a variety of use cases, depending on the boundaries. Window Calculations have different options including Average, Minimum, Maximum, Sum, Median, and Count. Tableau also has a variety of statistical options, covered in Recipe 16.4.

Problem

You want to bring back the window average across the worksheet, to replicate the average reference line value.

Solution

1. Start with continuous months on Rows and Sales on Columns:

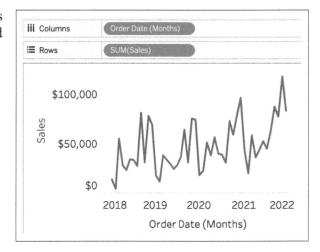

2. Add an average reference line to refer back to, and make sure we are getting the correct value back:

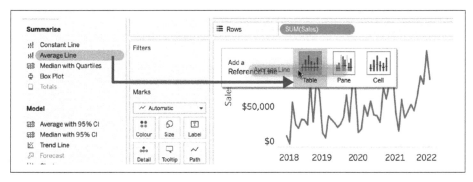

3. Creating a new calculation, on the right-hand side, search for Window Calculation, which gives you a description of what the calculation is doing:

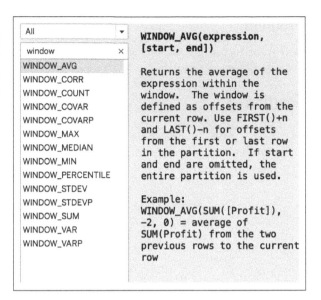

4. Double-click WINDOW_AVG and add SUM(Sales). The calculation assistance also gives you the option to add start and end boundaries, which is where you add values for Moving Average:

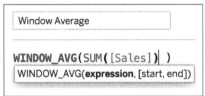

5. Finally, if we add this calculation to Detail and change the reference line to the Table Calculation, you'll see we have brought back the average value:

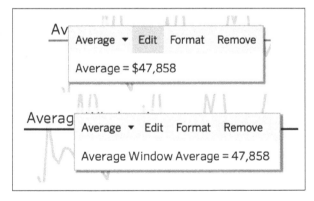

Discussion

Window Calculations allow you to use these in further analysis and calculations, to compare above and below the average. Another way to do this is with LOD calculations (see Chapter 17). Table Calculations are influenced by additional dimensions being added to a view, whereas you can specifically set the level of detail with an LOD calculation, so depending on what is in your view and what level you need to compute at, you will use Table Calculations or LOD calculations.

If we add any additional level of detail to the view (for example, Category), you might need to change how the Table Calculation is computing. The more details added to a chart, the more challenging it becomes to determine how a Table Calculation is calculating, or how to adjust it to get it to calculate the way you want it to.

Other than the Window Calculations already mentioned, statistical variables can also be calculated with a Window Calculation. That includes correlation, percentile, standard deviations, and variance. Use the search function in the calculation window for more information on how to use each one.

16.4 How to Create a Control Chart

A *control chart* helps you visualize whether a particular point is outside one or more standard deviations away from the mean. These can be used in many environments. The way to build them includes using Window Calculations mentioned in Recipe 16.3.

Problem

You want to create a control chart, to show which months are outside of one standard deviation away from the mean sum of sales.

Solution

1. Start by duplicating or re-creating Recipe 16.3. You will need the window average calculation.

2. Once again, you can add the Analytics Distribution Band to show the one standard deviation of the mean. To do this, go to the Analytics pane and drag Distribution Band onto the view:

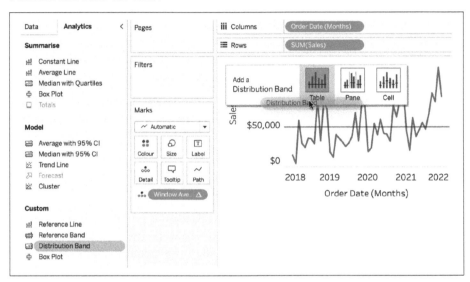

3. Select the Standard Deviation option:

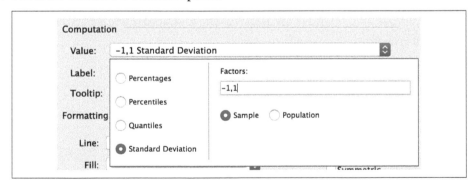

Your current view should now look like this:

This view shows the months that are above and below the standard deviation lines, but if you want to visually show the marks that are above and below the standard deviation lines, you will need to create the calculations.

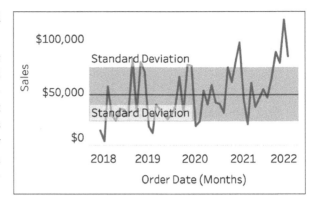

4. The first calculation is WINDOW_STDEV, which calculates the standard deviation across the view:

5. As we already have the window average, we can now create the standard deviation line above and below. To do this, we need to add or subtract the calculations:

6. These calculations replicate the standard deviation reference lines on the view. To sense check, add both calculations to Detail and check the values:

We have one last calculation to make this visually more appealing. We need to highlight which values are above or below the standard deviation, plus the average.

7. We can use a Boolean calculation here to ask if the sum of sales is greater than or equal to the plus standard deviation, or if the sum of sales is less than or equal to the minus standard deviation:

```
Above Below STD
─────────────────────────────
SUM([Sales])>=[+STD]
OR
SUM([Sales])<=[-STD]
```

You can also write this calculation as an IF statement to allow more than just True/False for the end user. That will look something like this:

```
IF SUM([Sales])>= [+STD] THEN "Higher than Normal Sales"
ELSEIF SUM([Sales])<= [-STD] THEN "Lower than Normal Sales"
END
```

8. To add this to the view, start by creating a second axis:

9. Add the new calculation to Color on the second Marks card and change to Circle:

10. Finally, create a Dual Axis and Synchronize. Also, hide your secondary axis, to make your view look like this:

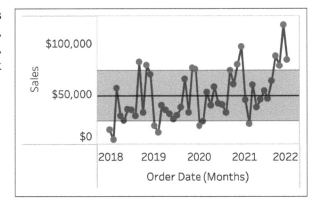

Discussion

A control chart enables users to statistically see whether the data has outliers. They can be parameterized by multiplying the standard deviation part of the calculation by the parameter, like this:

```
STD
```
Results are computed along Table (across).
```
WINDOW_STDEV(SUM([Sales]))*[Parameters].[STD]
```

To make full use of this parameterized version, I highly
recommend changing the reference lines to use the plus
and minus standard deviation
calculations you already
created:

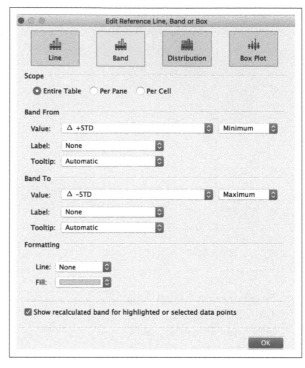

Finally, if you now want to change it to within two standard deviations of the mean,
you can change the parameter and it will update the reference line and the points outside the reference lines:

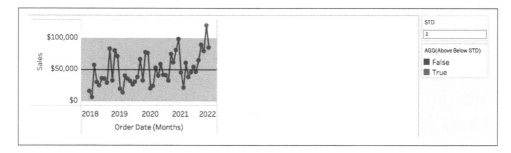

Summary

Table Calculations are very powerful in Tableau. Throughout the book, there have
been a variety of options to practice using Table Calculations; the majority are Quick
Table Calculations, but being able to understand how Tableau writes these as functions is incredibly important if you want to fully understand how Tableau works.

CHAPTER 17

Level-of-Detail Calculations

LODs are a very powerful feature of Tableau. Much like Table Calculations, they allow you to compute a value at a level of granularity that is different from the default visualization level of detail. However, the difference between Table Calculations and LODs is that Table Calculations are computed through dimensions or discrete fields that are in the view, whereas LODs can compute using fields that are not in the view or at a different level of granularity. You can use three types of LODs: Include, Exclude, and Fixed. Each has a different use case, which we will explore in this chapter.

When you want to create an LOD calculation, you can use the help guide inside of a calculation to understand each type of LOD. In the top of that calculation window, you will see how to formulate an LOD, and you will notice it has its own syntax:

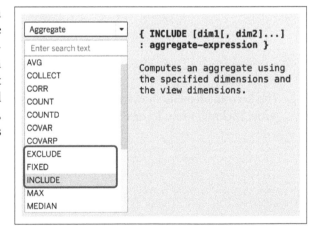

The difference between the three types of LODs is that an Exclude is less granular/more aggregated than the visualization level of detail; an Include adds additional granularity/less aggregation than the visualization level of detail; and finally, a Fixed LOD is at the level of granularity you specify regardless of what is in your view:

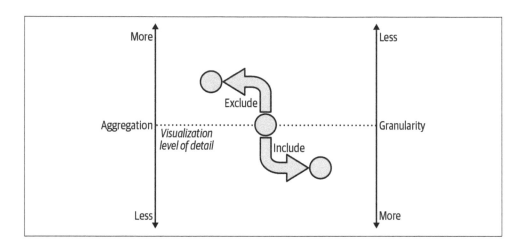

 Granularity is defined as "dimensions in the view."

Building an LOD calculation requires slightly different syntax from other calculations and is as follows:

1. The first element in an LOD always starts with an open curly bracket "{":

First LOD

{

2. The second element shows the type of expression you want to write: Include, Exclude, or Fixed:

First LOD

{INCLUDE

3. The third element, which is optional and depends on the use case with your calculation, is a comma that separates the list of dimensions that you want to compute the calculation with:

```
{INCLUDE [State]
```

4. The fourth element is a colon ":", separating the first half of the expression from the second half.

```
{INCLUDE [State] :
```

5. The fifth element is an aggregated value:

```
{INCLUDE [State] : SUM([Sales])
```

6. Finally, close off the expression with a closed curly bracket "}".

```
{INCLUDE [State] : SUM([Sales])}
```

Every LOD calculation you create will follow the same syntax. What you are trying to do with the calculation determines which dimensions and measures you add into the calculation.

The aggregated value could be either a measure already in the data, or it can be a calculation to return a specific part of a value, but it could also be a dimension or a date, as long as the fifth element is aggregated to either a Min, Max, Sum, Avg, CountD, or Count.

Creating the right LOD calculation could take many experiments to fully understand what the calculation is doing; however, the next four sections will hopefully give you insight into which expression type you will need to use.

17.1 Include LOD

Include LODs will contain any dimensions already in the view plus any extras you add into the third element, before the colon, in the calculation. An Include LOD will add more granulation and less aggregation.

Problem

You want to calculate the average sales per product for each subcategory, without including the product in the view.

Solution

1. Create a new calculation.

2. As the problem says, we don't want to show the product in the view. This means we need to "include" it into our LOD calculation:

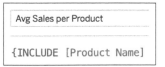

```
Avg Sales per Product

{INCLUDE [Product Name]
```

3. The second half in the calculation needs to be an aggregate measure.

4. For this example, we need to calculate the sum of sales for every product:

```
Avg Sales per Product

{INCLUDE [Product Name]    :SUM([Sales])}
```

5. Then take the average of those values, by wrapping the LOD in an average aggregation:

```
Avg Sales per Product

AVG(
{INCLUDE [Product Name]    :SUM([Sales])}
    )
```

LODs are considered nonaggregates, which allow you to aggregate their results.

6. To show how this calculation works, add Sub-Category to Rows and this new calculation to Columns:

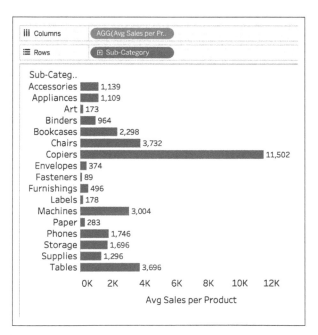

Discussion

To fully understand what this calculation is doing, it is important to understand the problem. This example shows the average sales for each product by subcategory.

Let's use the following view as an example for Accessories only. This view shows the sum of sales for every product within the subcategory of Accessories; we want the average of those values, which can be seen with the reference line:

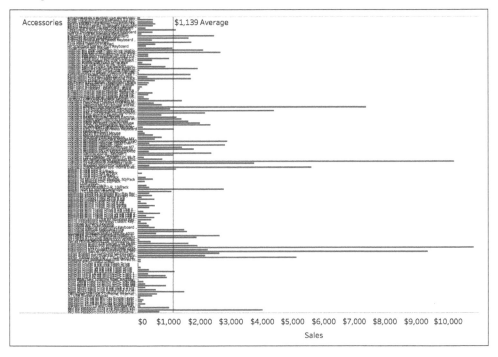

The reference line represents the average sum of sales across all the products, which is the same value being brought back by the LOD calculation we created in the recipe:

We need to create an LOD calculation to get this value back because we want the value to be at a certain granularity. If we compare the LOD

to just average sales, you can see the difference. Average Sales looks across the whole of the data, whereas the LOD is calculating the sum of sales for every product and then taking the average.

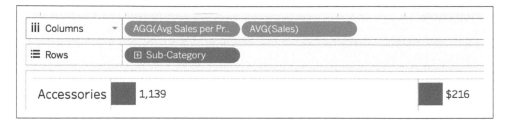

Accessories ▮ 1,139 ▮ $216

Tableau has an order of operations, which means different functions/expressions/calculations happen at a certain point. With Include and Exclude, calculations come after any filters in the view:

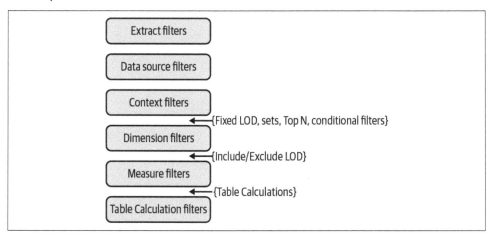

This means that if you add any filter to the view, it will happen before the calculation. If we filter to a particular order date, this will change the average sales per product value to that particular date:

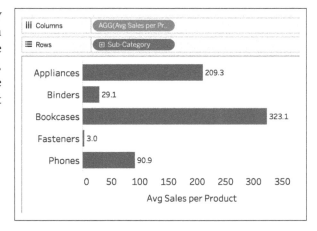

Your use case will determine whether you want to apply the filters to the LOD calculations. If you want to fix the value regardless of filters, take a look at Recipe 17.3, which goes through a Fixed LOD calculation.

17.2 Exclude LODs

The reverse of an Include LOD is the *Exclude LOD*, which removes granularity within your view. Like the Include expression, it uses all the dimensions in the view, but instead of including the extra dimensions mentioned in the calculation, it ignores them.

Problem

You want to calculate the total sales by region, to see the percentage contribution per state to that region.

Solution

1. To understand how this calculation is working, let's start with a text table with Region and State on Rows, and Sales on Text:

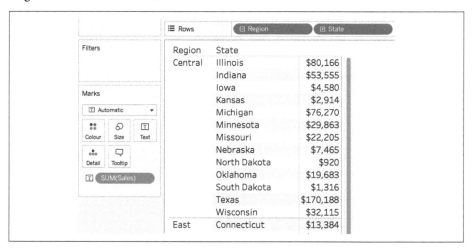

2. We want to calculate the sum of sales for each region. But because we want the view to remain at the region and state level, we have to create an Exclude LOD to create the total per region.

3. The syntax of an LOD has already been discussed in Recipe 17.1; the only difference is we are using Exclude instead of Include:

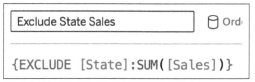

4. When we add this calculation to the view, you will see that you get the same value across every state in the same region:

Region	State	Sales	Exclude State Sales
Central	Illinois	$80,166	501,240
	Indiana	$53,555	501,240
	Iowa	$4,580	501,240
	Kansas	$2,914	501,240
	Michigan	$76,270	501,240
	Minnesota	$29,863	501,240
	Missouri	$22,205	501,240
	Nebraska	$7,465	501,240
	North Dakota	$920	501,240
	Oklahoma	$19,683	501,240
	South Dakota	$1,316	501,240
	Texas	$170,188	501,240
	Wisconsin	$32,115	501,240
East	Connecticut	$13,384	678,781
	Delaware	$27,451	678,781
	District of Columbia	$2,865	678,781
	Maine	$1,271	678,781
	Maryland	$23,706	678,781
	Massachusetts	$28,634	678,781
	New Hampshire	$7,293	678,781
	New Jersey	$35,764	678,781
	New York	$310,876	678,781

5. The final step is to calculate the contribution percentage:

Contribution Percentage

```
SUM([Sales])/SUM([Regional Sales])
```

Default Number Formatting

When you create a percentage calculation, it doesn't default to a percentage format; you have to change this using the default properties. Right-click the calculation from within the Data pane and choose Default Properties > Number Format:

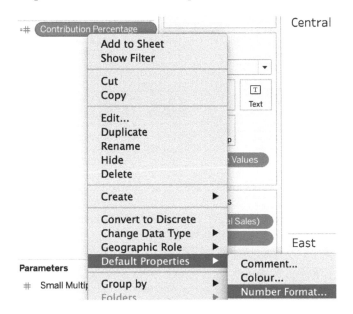

Then select percentage:

Changing the default number format means that whenever you use this field for the first time, it will always default to this format.

If you have changed the format of the field in a view, that will overwrite the default formatting.

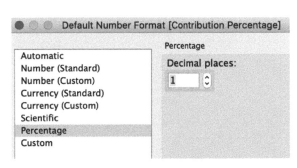

After adding the calculation to the view, the table then shows you the contribution of the state level sales to the overall region:

Region	State	Sales	Exclude State Sales	Contribution Percentage
Central	Illinois	$80,166	501,240	16.0%
	Indiana	$53,555	501,240	10.7%
	Iowa	$4,580	501,240	0.9%
	Kansas	$2,914	501,240	0.6%
	Michigan	$76,270	501,240	15.2%
	Minnesota	$29,863	501,240	6.0%
	Missouri	$22,205	501,240	4.4%
	Nebraska	$7,465	501,240	1.5%
	North Dakota	$920	501,240	0.2%
	Oklahoma	$19,683	501,240	3.9%
	South Dakota	$1,316	501,240	0.3%
	Texas	$170,188	501,240	34.0%
	Wisconsin	$32,115	501,240	6.4%
East	Connecticut	$13,384	678,781	2.0%
	Delaware	$27,451	678,781	4.0%
	District of Columbia	$2,865	678,781	0.4%
	Maine	$1,271	678,781	0.2%

6. If we add Sub-Totals to the view, you can see the contribution percentage adding to 100%:

Region	State	Sales	Exclude State Sales	Contribution Percentage
Central	Illinois	$80,166	501,240	16.0%
	Indiana	$53,555	501,240	10.7%
	Iowa	$4,580	501,240	0.9%
	Kansas	$2,914	501,240	0.6%
	Michigan	$76,270	501,240	15.2%
	Minnesota	$29,863	501,240	6.0%
	Missouri	$22,205	501,240	4.4%
	Nebraska	$7,465	501,240	1.5%
	North Dakota	$920	501,240	0.2%
	Oklahoma	$19,683	501,240	3.9%
	South Dakota	$1,316	501,240	0.3%
	Texas	$170,188	501,240	34.0%
	Wisconsin	$32,115	501,240	6.4%
	Total	$501,240	501,240	100.0%

Discussion

For this particular example, you might also consider using a Table Calculation computing pane down, but there are pros and cons to using an LOD over Table Calculations. First, you don't have to worry about the Compute Using element of this LOD as you do with a Table Calculation, because the Exclude will always exclude the dimension and then use the other dimensions in the view. Second, an LOD allows you to be explicit with the dimensions you use to aggregate the calculation. Table Calculations and Include/Exclude LODs do come after any filters are applied in Tableau. One type of LOD comes above most of Tableau's filters, and that's using a Fixed LOD.

17.3 Fixed LODs

Fixed LODs do exactly that: they fix the value and the precise level of detail that you specify into the calculation. They do not take into account any dimensions that are in your view—only those inside the calculation. Fixed LODs can give the same values as Include and Exclude LODs, except they don't change when a dimension filter is applied, nor do they change when any additional dimensions are added to the view.

Problem

You want to calculate the total regional sales to see the percent contribution by state, and then filter to show only Texas.

 When creating any type of calculation—Basic (Aggregation or Row-Level), LOD, or Table Calculation—I always start with a table. This allows you to see exactly what the calculations are doing and allows you to validate data before and during building visualizations.

Solution

1. Duplicate Recipe 17.2 or re-create the table.

2. Create a Fixed LOD at the region level for sum of sales:

Fixed Regional Sales

`{FIXED [Region]:SUM([Sales])}`

3. Add that to the current view:

Region	State	Sales	Exclude State Sales	Contribution Percentage	Fixed Regional Sales
Central	Illinois	$80,166	501,240	16.0%	501,240
	Indiana	$53,555	501,240	10.7%	501,240
	Iowa	$4,580	501,240	0.9%	501,240
	Kansas	$2,914	501,240	0.6%	501,240
	Michigan	$76,270	501,240	15.2%	501,240
	Minnesota	$29,863	501,240	6.0%	501,240
	Missouri	$22,205	501,240	4.4%	501,240
	Nebraska	$7,465	501,240	1.5%	501,240
	North Dakota	$920	501,240	0.2%	501,240
	Oklahoma	$19,683	501,240	3.9%	501,240
	South Dakota	$1,316	501,240	0.3%	501,240
	Texas	$170,188	501,240	34.0%	501,240
	Wisconsin	$32,115	501,240	6.4%	501,240
East	Connecticut	$13,384	678,781	2.0%	678,781
	Delaware	$27,451	678,781	4.0%	678,781
	District of Columbia	$2,865	678,781	0.4%	678,781
	Maine	$1,271	678,781	0.2%	678,781
	Maryland	$23,706	678,781	3.5%	678,781

4. Create another contribution calculation using the new fixed regional sales calculation:

Fixed State Contribution

```
SUM([Sales])/SUM([Fixed Regional Sales])
```

5. Add that to the view:

Region	State	Sales	Exclude State Sales	Contribution Percentage	Fixed Regional Sales	Fixed State Contribution
Central	Illinois	$80,166	501,240	16.0%	501,240	16.0%
	Indiana	$53,555	501,240	10.7%	501,240	10.7%
	Iowa	$4,580	501,240	0.9%	501,240	0.9%
	Kansas	$2,914	501,240	0.6%	501,240	0.6%
	Michigan	$76,270	501,240	15.2%	501,240	15.2%
	Minnesota	$29,863	501,240	6.0%	501,240	6.0%
	Missouri	$22,205	501,240	4.4%	501,240	4.4%
	Nebraska	$7,465	501,240	1.5%	501,240	1.5%
	North Dakota	$920	501,240	0.2%	501,240	0.2%
	Oklahoma	$19,683	501,240	3.9%	501,240	3.9%
	South Dakota	$1,316	501,240	0.3%	501,240	0.3%
	Texas	$170,188	501,240	34.0%	501,240	34.0%
	Wisconsin	$32,115	501,240	6.4%	501,240	6.4%
East	Connecticut	$13,384	678,781	2.0%	678,781	2.0%
	Delaware	$27,451	678,781	4.0%	678,781	4.0%
	District of Columbia	$2,865	678,781	0.4%	678,781	0.4%
	Maine	$1,271	678,781	0.2%	678,781	0.2%
	Maryland	$23,706	678,781	3.5%	678,781	3.5%

6. You will notice that currently Exclude and Fixed are giving back the same values from two different calculations. However, the effect of one over the other comes into play when we add a filter to just Texas:

Region	State	Sales	Exclude State Sales	Contribution Percentage	Fixed Regional Sales	Fixed State Contribution
Central	Texas	$170,188	170,188	100.0%	501,240	34.0%

Discussion

In the last step of this recipe, we use the Dimension filter to keep only Texas values. The calculations on the left are using the Exclude LOD, which comes after any filters. The calculations on the right are using the Fixed calculation and they remain unchanged. This is because of Tableau's order of operations. If you need some filters to apply to the fixed LOD calculation, you will need to add them to context, which moves it up in Tableau's order of operations.

Prior to filtering, here is the view that is looking at just Texas:

Region	State	Sales	Fixed Regional Sales	Fixed State Contribution
Central	Texas	$170,188	501,240	34.0%

For example, if you want to see the values for only the latest year of data, add Order Date to Filters, select the year, and then checkmark the box at the bottom:

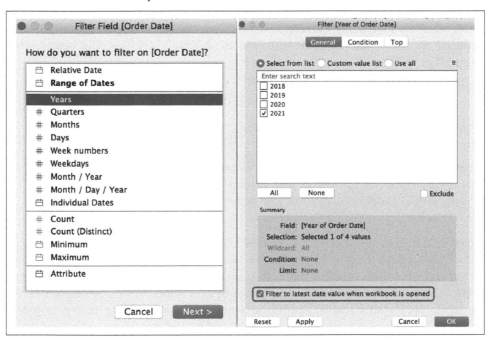

Notice the view has changed the contribution percentage because it's based on the latest year's sales ($43,422) instead

Region	State	Sales	Fixed Regional Sales	Fixed State Contribution
Central	Texas	$43,422	501,240	8.7%

of all years' sales ($170,188). It hasn't changed the Fixed Regional Sales, therefore resulting in a lower percentage.

To make sure this particular filter gets applied before the LOD calculation, you need to right-click the filter and select "Add to Context":

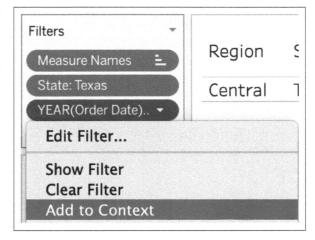

Now, when we look at the current view, you'll see the Fixed Regional Sales has been reduced to reflect the current year, which ultimately changes the contribution percentage:

Region	State	Sales	Fixed Regional Sales	Fixed State Contribution
Central	Texas	$43,422	147,098	29.5%

Fixed LODs don't always need to have dimensions before the colon. These are known as *table-scoped LODs*. When you remove a dimension, you can also remove the FIXED and the colon:

Fixed LODs have also been used to deal with duplicated data. The duplication could have been due to a join exploding the number of rows, or, in general, your data contains duplicates, and you want to keep only the maximum value. This is where knowing the details of your data is incredibly important, and also understanding what a row is. The introduction of the logical layer and relationships might reduce the need to remove duplicate data due to a join.

There is also a quick way of creating Fixed LODs. Using Quick LODs are great ways for getting started with LODs. To create one quickly, select the measure you want to use, press Ctrl (Command on Mac), and drag it on top of the dimension you want to use in the Data pane:

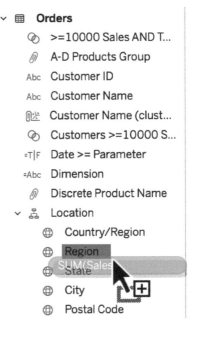

This creates a calculated field that is named using the measure and dimension in brackets:

Then when you look at the calculation, you can see what Tableau has created:

```
Sales (Region)                          Orders & Customer Details

{ FIXED [Region]: SUM([Sales]) }
```

17.4 Nested LODs

Many calculations give you the option of being able to use nested calculations, which you can also do with LODs.

Problem

You want to calculate the sum of sales per state, and then return the maximum value for each region. This will allow you to either find the highest state based on sales or compare the other states to the highest.

Solution

1. Once again, start with a table of region, state, and sales.

2. To begin building this nested calculation, bring back the sum of sales for each state:

Max State Sales by Region

```
{FIXED [State]:SUM([Sales])}
```

3. When we add this calculation to the view, you'll notice it is bringing back the same as the sum of sales. That's because the FIXED LOD currently matches the visualization level of detail:

Region	State	Sales	Max State Sales by Region
Central	Illinois	$80,166	80,166
	Indiana	$53,555	53,555
	Iowa	$4,580	4,580
	Kansas	$2,914	2,914
	Michigan	$76,270	76,270
	Minnesota	$29,863	29,863
	Missouri	$22,205	22,205
	Nebraska	$7,465	7,465
	North Dakota	$920	920
	Oklahoma	$19,683	19,683
	South Dakota	$1,316	1,316
	Texas	$170,188	170,188
	Wisconsin	$32,115	32,115
East	Connecticut	$13,384	13,384
	Delaware	$27,451	27,451
	District of Columbia	$2,865	2,865
	Maine	$1,271	1,271
	Maryland	$23,706	23,706
	Massachusetts	$28,634	28,634
	New Hampshire	$7,293	7,293
	New Jersey	$35,764	35,764
	New York	$310,876	310,876
	Ohio	$78,258	78,258
	Pennsylvania	$116,512	116,512
	Rhode Island	$22,628	22,628
	Vermont	$8,929	8,929
	West Virginia	$1,210	1,210

Now we need to find the maximum value from all those states, which means we bring back the value for Texas in the central region.

4. This is where we can nest LOD calculations. Edit your "Max State Sales by Region" calculation and add a FIXED at Region for the max of the inner LOD. Don't forget to close the maximum aggregation and the outer LOD:

Max State Sales by Region

```
{FIXED [Region] : MAX(  {FIXED [State]:SUM([Sales])}  )}
        Outer LOD               Inner LOD
```

5. Once we save the calculation, you will see that it brings back the value for Texas for every state in the central region, and the highest value for the other regions:

6. Calculating the maximum value allows you to compare the individual states to this maximum.

Region	State	Sales	Max State Sales by Region
Central	Illinois	$80,166	170,188
	Indiana	$53,555	170,188
	Iowa	$4,580	170,188
	Kansas	$2,914	170,188
	Michigan	$76,270	170,188
	Minnesota	$29,863	170,188
	Missouri	$22,205	170,188
	Nebraska	$7,465	170,188
	North Dakota	$920	170,188
	Oklahoma	$19,683	170,188
	South Dakota	$1,316	170,188
	Texas	$170,188	170,188
	Wisconsin	$32,115	170,188

Discussion

Nested LODs are very powerful when done correctly because they allow you to roll up the data in a predictable, intuitive, and structured way.

Other types of examples include comparing against the average sum of sales by region. This would require the following calculation:

```
Avg State Sales by Region

{FIXED [Region] : AVG(   {FIXED [State]:SUM([Sales])}   )}
```

The inner part of the nested LOD is always where to begin thinking about whether you need to create a nested calculation, to make sure the inner part is bringing back the correct value first. Once you have that, you can work on the outer part of the LOD. There are no limits on how many LODs you can nest; however, make sure you are referring back to the question you are trying to answer and the ability for someone else to understand what the calculation is actually doing. In my experience, I have never used any more than two levels of nested calculations.

You can also create your inner LOD as a separate calculation and use that as a reference inside your secondary calculation.

Using Quick LODs

If you know you need to create a nested LOD, you can also create this using quick LODs.

Start by creating the inner LOD by pressing Ctrl and dragging Sales on top of State in the Data pane to create the first LOD:

That creates a new calculation called Sales (State). We know we need the maximum value by region for our outer LOD. When creating a quick LOD, it takes the default aggregation into account. We can change Sales (State) default aggregation to maximum to help build our calculation. To do that, right-click the field and choose Default Properties, > Aggregation > Maximum:

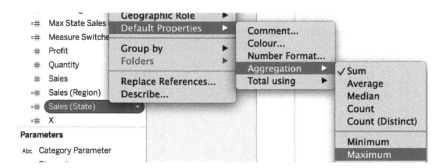

Now to create the nested LOD. You can press Ctrl and drag the Sales (State) on top of the Region dimension in the Data pane:

This calculation will now be called Sales (State) (Region), indicating a nested LOD. If we edit this calculation, you will see what Tableau has created:

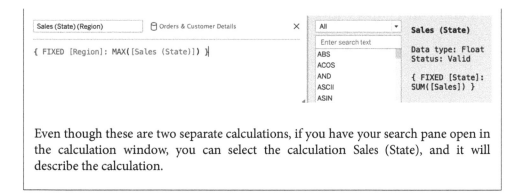

Even though these are two separate calculations, if you have your search pane open in the calculation window, you can select the calculation Sales (State), and it will describe the calculation.

Summary

Throughout this chapter, we have gone through the different types of level-of-detail calculations you can create, including how to create quick LODs. LODs are very powerful in being specific about the aggregation and granularity of values. You can also convert many Table Calculations into LODs if you need to get around Tableau's order of operations.

CHAPTER 18

Advanced Mapping

In Chapter 6, we covered basic mapping functionality within Tableau. This chapter covers the more advanced mapping techniques and functionality that you can do within Tableau. These advanced mapping techniques came into Tableau from 2018 onward and made an impression on how you can advance your mapping capabilities.

Always map responsibly and make sure you are taking into consideration proportionality (not creating a map that misrepresents the data).

This chapter uses United States mainland flights data to enable full use of the mapping capabilities.

 Tableau has some mapping capabilities built in—for example, the ability to select a state or city as a geographic field. To use the advanced mapping functionality, you will need to have a dedicated latitude and longitude field to use within spatial calculations.

18.1 Density Marks

The Density mark type was first mentioned in Chapter 11, where you learned to use this mark type on a scatter plot to visualize the highest concentration of data. This mark type is also appropriate for mapping. It allows the users to see the concentration of data around a given area.

Problem

You want to see the concentration of airports on the mainland US.

Solution

1. Double-click Origin Latitude and Origin Longitude, and add Origin Name to Detail:

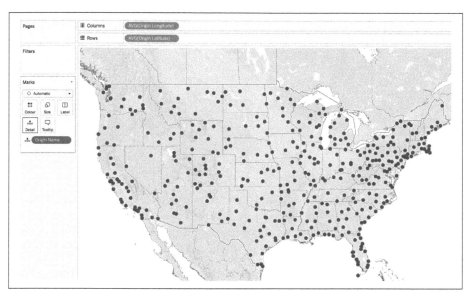

2. Change the mark type to Density:

Your view will now look like this:

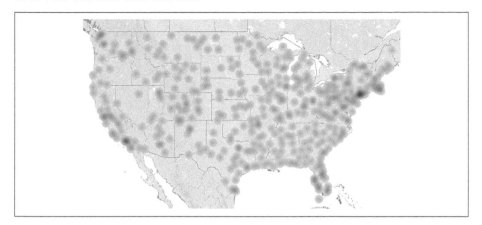

3. With the Density mark type, Tableau has created new color palettes to use. To change the Density color, click Color on the Marks card and edit the color:

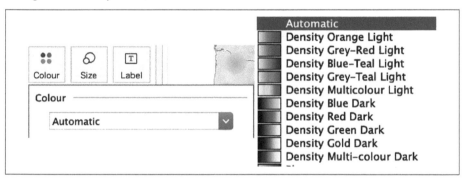

4. Finally, for this particular view, change the size of the mark:

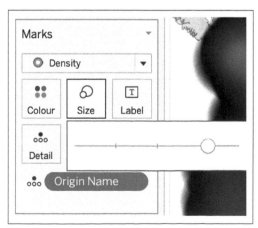

Your view should now look like this:

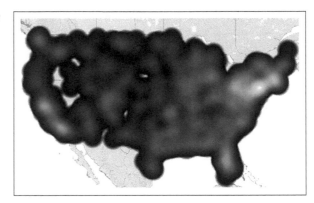

Discussion

Using the Density mark type shows you the highest concentration within a particular view. This example shows that the highest concentration of airports is around the Northeast coast and within California.

For more information about the Density mark, refer to Recipe 11.3.

18.2 MakePoint

To perform any of the spatial calculations, Tableau needs a specific spatial object. A *spatial object* can be something that is already within a spatial file or can be generated within Tableau as long as you have a latitude and longitude or *x* and *y* coordinates. They come in the form of points, lines, and polygons. To start any spatial calculations, Tableau needs a specific point object; this is where the MakePoint calculation makes an appearance.

Problem

You need to create a spatial point of origin for airports to enable further spatial analysis.

Solution

1. Start by creating a calculated field and type `MakePoint`:

2. For MakePoint to work, you either need latitude and longitude or *x* and *y* coordinates. Add the origin latitude and longitude into your calculation:

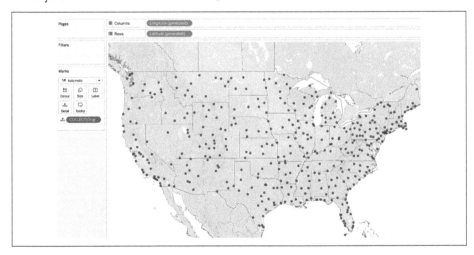

MAKEPOINT([Origin Latitude],[Origin Longitude])

3. Once you have saved this calculation, double-click it to add it to the view:

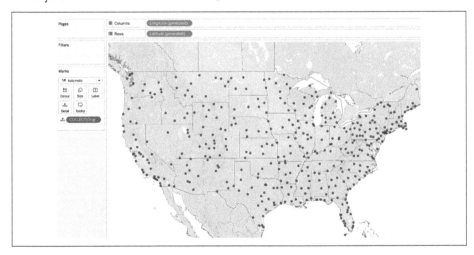

Discussion

When using MakePoint in the view, you will notice a few differences compared to using the origin latitude and longitude in Recipe 18.1. First, we are now using the generated latitude and longitude instead of origin latitude and longitude. This is because Tableau has now created its own latitude and longitude for the origin "airports":

Second, on the Marks card, you have a new field that starts with Collect. Collect is specific to any spatial object that creates the generated latitude and longitude:

With this Collect field on the Marks card, when you hover over a mark on the map, you might notice that all the marks highlight together, and no tooltip appears to say which airport the individual point belongs to:

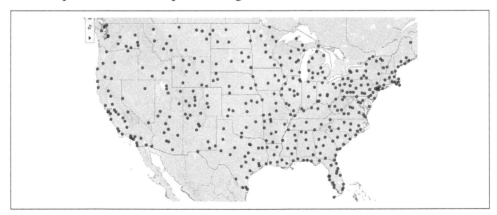

To break up this view, we need to add Origin Name to the Marks card:

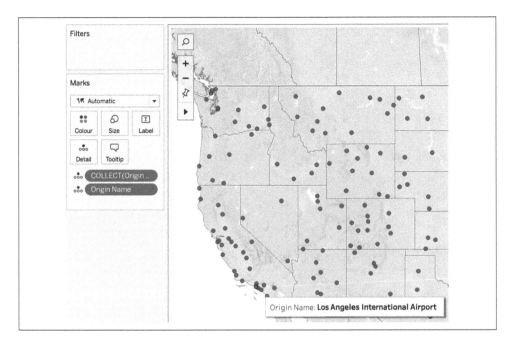

Origin Name: **Los Angeles International Airport**

This particular example gives you all of the origin airports, but what if we want to see where those airports fly to and then connect them with a line?

18.3 MakeLine

A *hub and spoke chart* shows a connection between a start and end point on a map. For this data set, it is the connection between an origin airport and a destination airport, a Route.

MakeLine is Tableau's answer to creating this chart type.

Problem

You want to see the connections between each airport using a hub and spoke map.

Solution

For this calculation to work, you need two spatial points to plot. The origin point was created in Recipe 18.2.

1. Create a destination point using the MakePoint calculation again:

2. Create a new calculation and start typing `MakeLine`:

3. MakeLine takes a start point and an endpoint and creates a spatial line object.

4. Add the origin point and destination point into this calculation:

5. Double-click this field to create the following view:

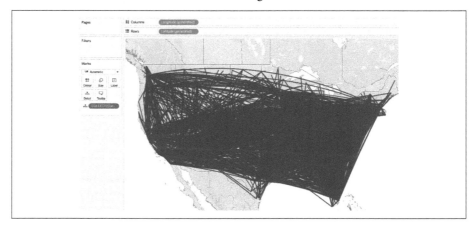

6. Next, we filter to a single origin airport to show exactly what Tableau has done. I'm filtering to JFK:

This is what your current view should look like:

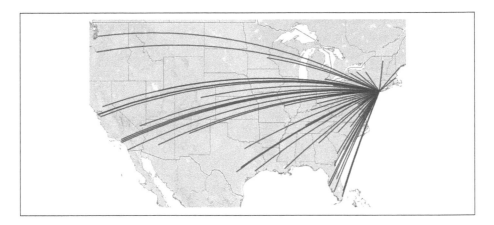

7. But as in Recipe 18.2, if you hover over the minute, all the connections are as one:

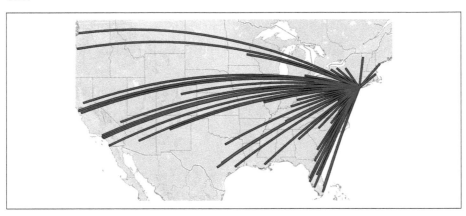

8. To break down the view, add the field's origin Name and Destination Name to Detail:

9. Now, hover over a single line, and you should see the origin airport and destination airport name:

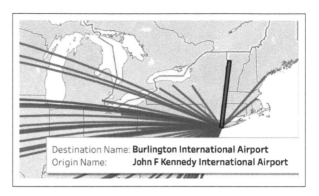

Destination Name: **Burlington International Airport**
Origin Name: **John F Kennedy International Airport**

Discussion

This type of chart allows you to show origin and destination connections or route mapping. The MakeLine calculation also considers the curvature of the world; that's why these lines are not straight but slightly arched.

Other examples for using a hub and spoke map would be trade direction and locations.

What if you wanted to add circles as endpoints to each one of the lines? That's where Marks Layer will help.

18.4 Marks Layer

Marks Layer does exactly what its name says: it builds layers of spatial data on top of each other. The number of marks layers is infinite; however, the more layers and more data you are displaying might affect the performance of the visualization.

> ### Top Tip 9: Marks Layers
>
> Mark layers have the ability to change the way you work with Maps within Tableau. Being able to layer multiple map layers together onto one map is powerful.

Problem

You want to display start and endpoints on each route from the hub and spoke map.

Solution

1. Duplicate or re-create Recipe 18.3.

2. Starting with the Origin Point, drag and drop the calculation Origin Point on top of the Marks Layer option that appears:

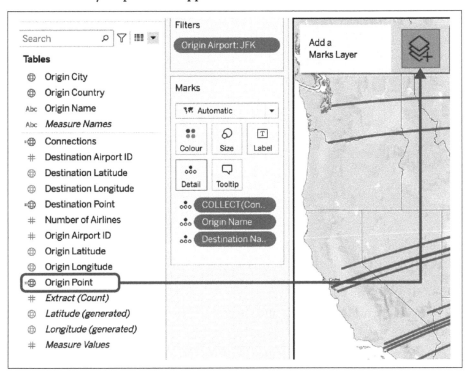

This creates two Marks cards on the left side:

3. Change the Origin Point Marks card to the Square mark type and change the Color to black:

The point of origin will now be a black square:

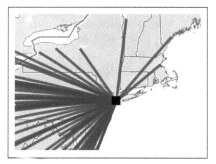

4. To add the endpoint, drag and drop the Destination Point onto a new Marks Layer:

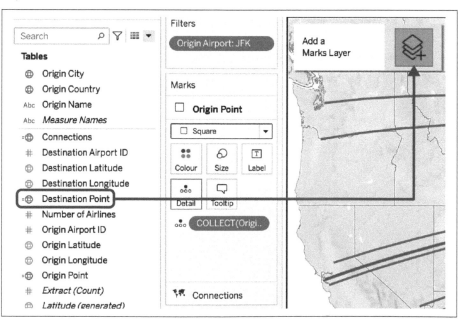

5. This adds another Marks card on the left, and we want to change this to Circle, the Color to black, and add Destination Name to Detail to split the marks to the correct level of detail:

Your final view will look like this:

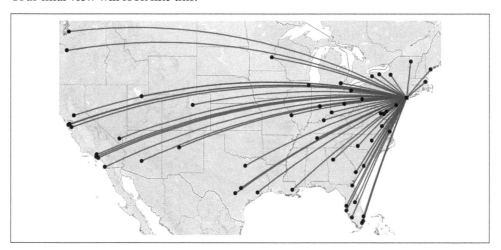

Discussion

Marks Layer allows you to layer many spatial objects on top of each other. The key differences with the Marks Layer Marks card compared to any non-map Marks card is that you can quickly reorder the Marks Layers with drag-and-drop; moving a Marks Layer to the top means that will be the front element on the map:

Notice that the connections (lines) are now in front of the origin and destination points:

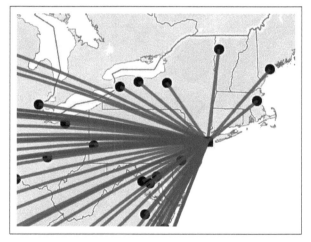

The second difference between a Marks Layer Marks card and a non-mark layer Marks card is you can rename a particular layer to allow for a better understanding when someone else has to use your workbook. To rename, you can right-click the header of a Marks card and select Rename:

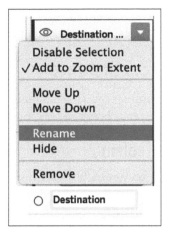

Another difference is the ability to disable the selection on the Marks Layer. That means you won't have any tooltips, nor will you be able to select a specific mark on that layer. To do this, you can right-click a layer head and select Disable Selection:

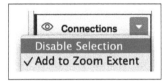

Now, when you try to hover or select a line, you won't be able to.

Finally, you also have the ability to hide a Marks Layer. The reason you might hide a layer is that you're still in the development phase, data might not be correct, or an update has created an error and therefore you can hide the layer.

Layer Control

Your users also have the ability to control what mark layers they want to see and interact with. This is called *Layer Control*. You can check and uncheck layers and also lock them to disable interaction:

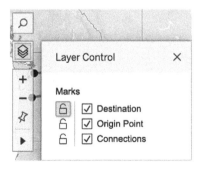

Prior to Tableau 2020.4, Marks Layer was not a feature and Tableau could handle only dual-axis maps, meaning only two layers on a map. See Recipe 6.3.

18.5 Distance

Another spatial technique that has been added to Tableau is the ability to calculate the distance between two points.

Problem

You want to know the distance between the origin and destination airports.

Solution

1. Create a new calculation and start typing **DISTANCE**:

2. Within this spatial calculation, you need to use spatial points, so if you haven't created the origin and destination point, you will need them. See Recipe 18.2.

3. Add the origin and destination point into the calculation:

4. The last part of this calculation is defining the metric of distance you want to calculate; in this case, we are going to use miles. You will need to add **mi** to the end of the calculation:

5. Now, when we add this to Color on our Connections Marks card, you can see the distance of all the connections, and it is added to the tooltip of the line:

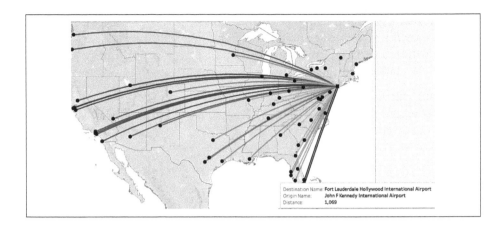

Destination Name: **Fort Lauderdale Hollywood International Airport**
Origin Name: **John F Kennedy International Airport**
Distance: **1,069**

Discussion

Being able to calculate distance within Tableau can be incredibly useful, especially if you want to filter to a specific distance away from a mark.

Within the distance calculation, you are asked to specify what measure you want to use to calculate the distance. You can add four options to this: "km" for Kilometers, "mi" for Miles, "m" for Meters and "ft" for Feet. Your use case will determine which one you want to use. You can also allow your user to specify the measure you use, by switching the last part of the calculation with a parameter with the four options. Your parameter settings and calculation would need to look something like this:

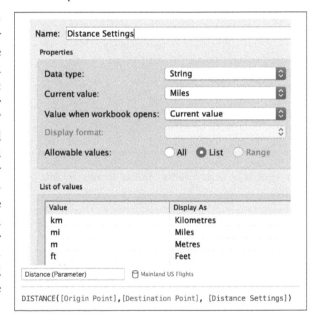

This gives your user the ability to change the distance settings.

You can also use this distance calculation to give the user the ability to see only destination airports within a certain distance by creating a Boolean calculation, like this:

Distance <=500mi

[Distance]<=500

When this calculation is added to Filters and True is selected, the view will show only destination airports within five hundred miles of the origin:

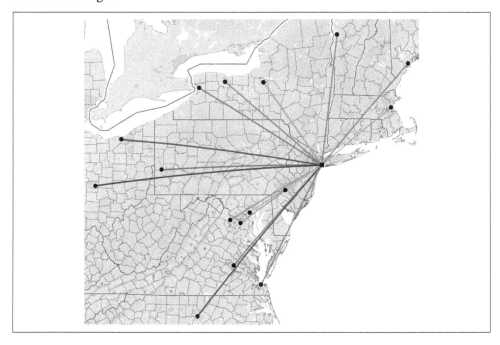

Another way to create the ability to see the airports within a certain distance is using Buffer.

18.6 Buffer

Buffer calculations create a set radius around a point, creating a circle around the mark.

Problem

Create a buffer around an origin airport.

Solution

1. Create a new calculation and type **Buffer**:

2. As with the other spatial calculations, we need to use the spatial point for which we want the buffer to surround. Therefore, you should add Origin Point to the calculation:

3. Next, we need to add how far of a radius around the origin point we want to create the buffer:

4. Finally, add the measure for distance; in this case, it is miles:

Buffer ◷ Mainla

BUFFER([Origin Point],500,'mi')

5. Now that we have the calculation, you can drag and drop it to create a new map layer:

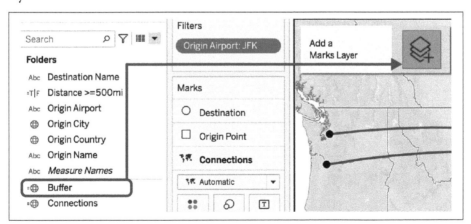

Your view will now look like this:

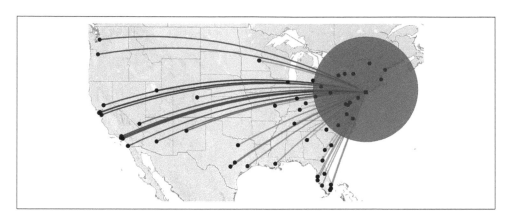

Discussion

Buffer is used to create a radius around a specific point. With Marks Layer, you can change the order, so the buffer is behind the main marks:

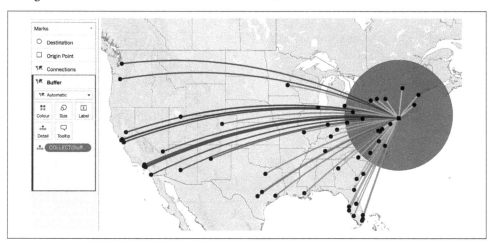

However, if you wanted to keep only those specific values, you would have to use it as part of a spatial join process.

18.7 Spatial Joins

Spatial joins are specifically used when the data contains spatial information: a point, polygon, or line. You can tell Tableau how you want them to join, and the unique join type with a spatial join is the option to select Intersects.

Problem

You want to keep only the airports within 500 miles of the origin airport.

Solution

1. Go back to the Data source pane and click the drop-down to return to the physical layer:

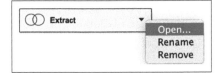

2. Within this current join, you can see it is just joining on Destination Airport ID:

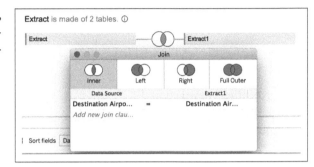

3. We want to add a new join clause, and because at this level it doesn't recognize the spatial calculations we have already created, we need a new join calculation:

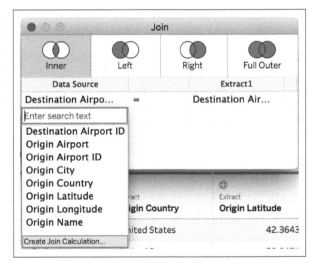

4. To start with, use the MAKEPOINT function, to give the origin the spatial point:

5. We want the buffer to be around the origin airport, so we need to add the buffer calculation to this:

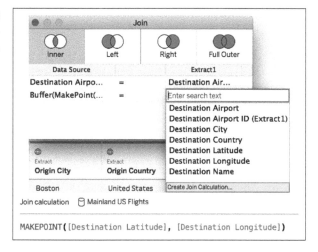

Join calculation ⊟ Mainland US Flights
```
Buffer(
MakePoint([Origin Latitude],[Origin Longitude]),
500, 'mi'
)|
```

Notice that I have added new lines and spaces within the calculation. This is because Tableau doesn't account for them, meaning you can space them out to make them easier to read and understand.

6. Give the destination a spatial point, too, by creating another join calculation on the right side:

Join

Inner   Left   Right   Full Outer

Data Source		Extract1
Destination Airpo...	=	Destination Air...
Buffer(MakePoint(...	=	Enter search text

Destination Airport
Destination Airport ID (Extract1)
Destination City
Destination Country
Destination Latitude
Destination Longitude
Destination Name

Create Join Calculation...

Extract	Extract
Origin City	**Origin Country**
Boston	United States

Join calculation ⊟ Mainland US Flights

```
MAKEPOINT([Destination Latitude], [Destination Longitude])
```

7. Once we click off this, Tableau will look as if it has created an error, and that's because we are trying to do a normal join type on spatial data. To rectify this, we can change the join type in the middle to Intersects:

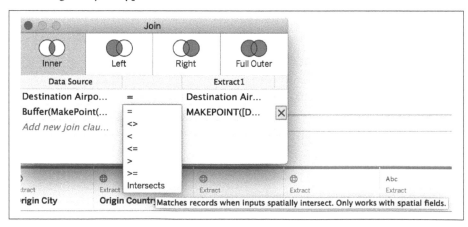

Join

Inner   Left   Right   Full Outer

Data Source		Extract1
Destination Airpo...	=	Destination Air...
Buffer(MakePoint(...	=	MAKEPOINT([D... ✕
Add new join clau...		

=
<>
<
<=
>
>=
Intersects

tract	Extract		Extract	Extract	Abc Extract
rigin City	Origin Countr				

Matches records when inputs spatially intersect. Only works with spatial fields.

8. Now when we go back to the worksheet, you will see that Tableau keeps only those destination airports within the buffer zone:

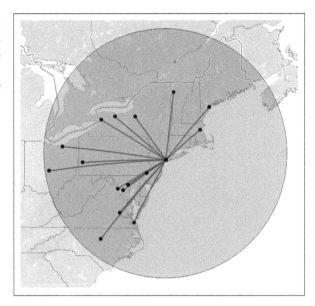

Discussion

Spatial joins are used for specific spatial file types (.shp, .kml, etc.).

This particular example is good for when you want to zero in on a specific buffer zone. However, what if you want the ability to change the buffer area? If you change the calculation here, it won't change the data being brought through because we have used a specific buffer zone within the join.

If you want to keep all your data, refer back to using the Boolean distance calculation mentioned in Recipe 18.5.

Summary

This chapter has gone through advanced techniques of working with maps in Tableau. You should now be able to create geographic fields by using MakePoint and MakeLine. Having the ability to calculate distance and use buffers adds additional functionality to your maps.

You can also create stadium and building maps if you have the data. Ryan Sleeper, in *Practical Tableau*, walks through a fantastic example in Tip 35, "How to Make Custom Polygon Maps."

CHAPTER 19

Advanced Dashboarding

The basics of dashboards were covered in Chapter 7, which included how to build a basic dashboard, basic interactivity, basic objects, and layouts. This chapter covers more advanced techniques when creating your dashboards, to give you more flexibility to create the right user experience for your audience.

19.1 Actions

You can perform several types of actions on a dashboard. Filtering (Chapter 7), set actions (Chapter 14), and parameter actions (Chapter 15) have all previously been mentioned, but a few more action types allow further interactivity.

Highlight Action

A *highlight action* allows you to select a mark and, if the relevant field is available in other charts, it will show you that mark.

Problem

When selecting a week, weekday, or both, you want to highlight the relevant week or weekday in the marginal histogram:

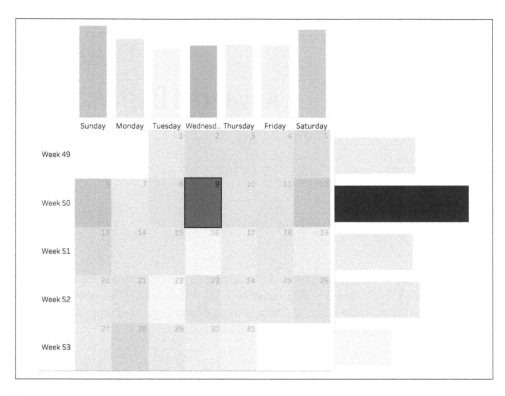

Solution

This dashboard is from Recipe 8.3.

1. On the dashboard, choose Dashboard > Actions:

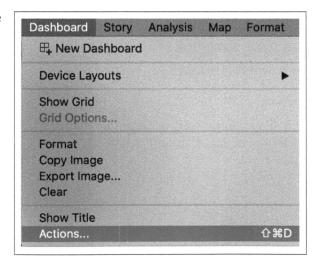

2. Add a new action and add a Highlight action:

This will give you a pop-out box similar to the other actions you have created.
Nothing needs to be changed inside this action:

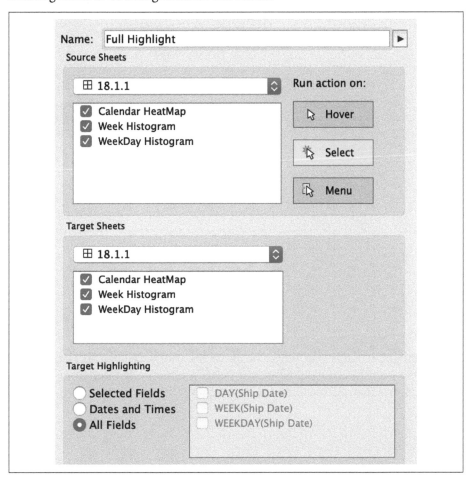

3. Click OK to close the action box. When you click a square on the heatmap for weekday and week, Tableau will highlight the respective weekday and week on the bar charts:

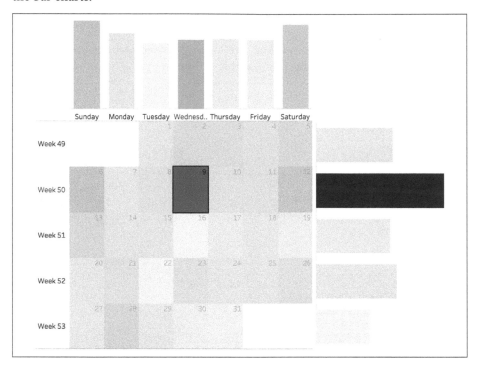

Discussion

This highlight action is very powerful across the whole dashboard. The way we have currently set up the action allows you to select a weekday or week from the marginal bar charts, and that will highlight the heatmap as well:

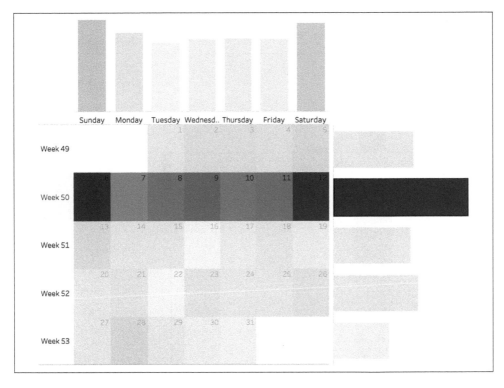

Unfortunately, you cannot change the default yellow highlighting that Tableau uses.

One of the features within the highlight action is the ability to select which fields you want to highlight. This is needed because you might have a lower level of detail in your visualizations, but you only want to highlight the higher level. This can be done at the bottom of the highlight action:

For example, choose Selected Fields and select WEEK-DAY(Ship Date):

When you close the action box, your highlight will now only highlight weekdays instead of week and weekday:

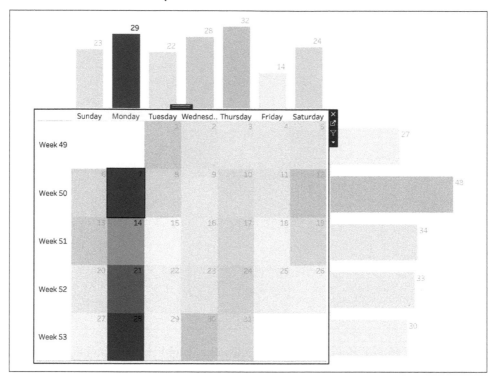

URL Action

A *URL action* allows you to select a mark and go to a specific web page in a browser or use the Web Page object ("Web Page" on page 622).

Problem

You want your user to select a state on a map and go to the Wikipedia page associated with that state.

Solution

1. Add a state-based map onto a dashboard:

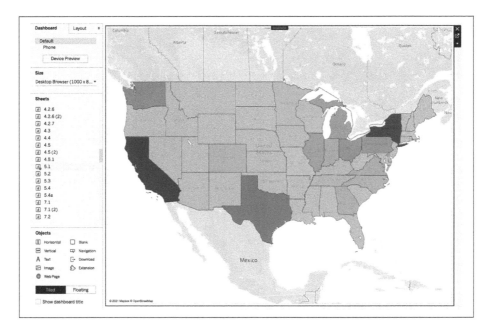

2. Choose Dashboard > Actions:

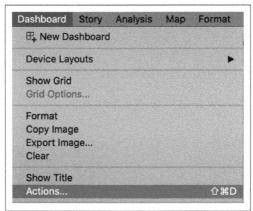

3. This time, select "Go to URL":

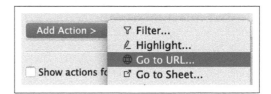

The pop-up action box is similar to other actions, except that the bottom half is different:

5. We are going to do a Menu action this time, which means it adds a hyperlink to the tooltip upon selection, so when we select the state, we want to tell the user which Wikipedia page we are going to for that state. We can do this by adding the State field to the name of the action:

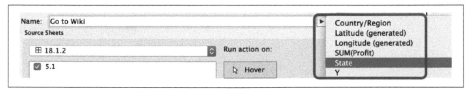

This now makes the name of the action look like this:

6. Add the base URL to the URL section of the action:

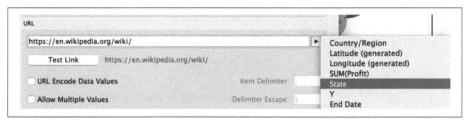

7. Again, we need to add the State field to the end of this URL, so Tableau knows which state Wikipedia page you would like to view:

To find the correct URL, I looked up one state to copy the syntax of the Wikipedia URL. This also identified where to put the dimension in the URL.

This will make your URL link look like this:

8. Select a state on the map, and within the Tooltip the action will pop up:

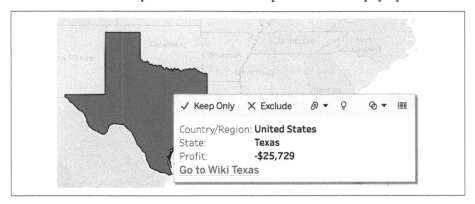

And then a selected browser web page will pop up, taking you to that URL:

Discussion

When using URL actions, it is important that the URL link you provide will take you to the exact page that you want it to. For this example, we are selecting the state of New York:

Wikipedia gets an error, because it doesn't specifically know which New York page you are trying to access:

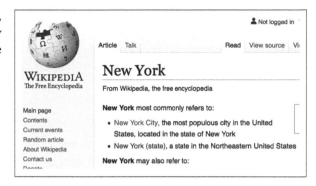

So, for this specific example, we would need to add **_(state)** to the end of the URL link:

Now when we select the state of New York, it will take you to that specific state's Wikipedia page:

See "Web Page" on page 622 for how to use the URL actions with a Web Page object.

Other use cases for using URL links include having a unique identifier for a customer or opportunity that can then link to a custom CRM or Salesforce instance.

19.2 Dashboard Objects

Like actions, several dashboard objects have been mentioned in Chapter 7. This section will cover the more advanced dashboard objects which enhance your user's experience.

Web Page

Tableau has the ability to embed a web page inside your Tableau dashboards. This is incredibly useful when using the URL actions mentioned in "URL Action" on page 616, as it stops the URL from going to the browser and instead uses the embedded web page.

Problem

You want to embed a web page inside a Tableau dashboard, which allows your users to select a state and go to that Wikipedia page.

Solution

1. Continuing with the dashboard used in "Highlight Action" on page 611, drag and drop a Web Page object onto the canvas:

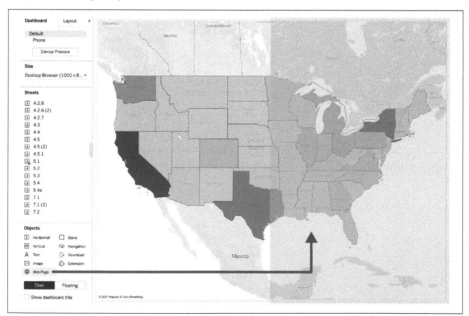

This will create a pop-up box, allowing you to add the URL:

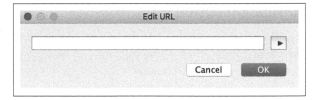

2. You have the option to leave this blank or add your URL. But because we already have a URL action, you can leave this blank.

3. Now when you click a state, the web object will populate with the URL from the action:

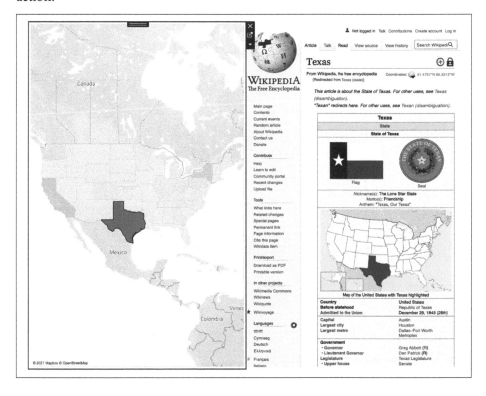

Discussion

Embedding a web page inside your dashboard is useful if you have a web page you need to link out to. However, be aware that it does need to have ftp, http, or https prefixed and, for security reasons, other protocols and UNC paths are not supported.

Also, when publishing to Tableau Server, you will need to make sure your Tableau Server has access to the internet to open the web page in the dashboard.

Navigation Button

Another dashboard object is the *navigation button*, which allows you to go from one dashboard to another. These buttons can be used as forward or backward navigation through a series of dashboards.

Problem

You want your user to go to different dashboards by using buttons.

Solution

1. On any dashboard, drag and drop the Navigation object onto the canvas:

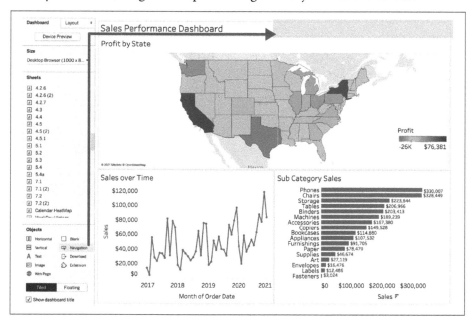

2. To tell Tableau what to do with this object, you need to click the drop-down by the object and select Edit Button:

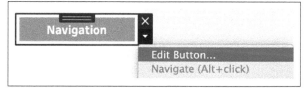

This will give you a pop-up box:

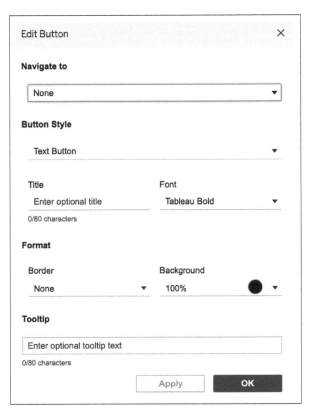

3. We need to tell the button where we want to navigate to; using the first drop-down, we can select a different dashboard:

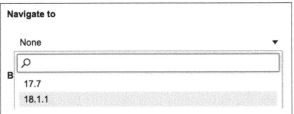

4. Choose what style of button you would like. You can select an Image or Text Button:

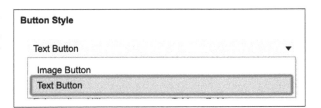

5. When sticking with the text button, you can choose what the button says using the Title section:

6. Finally, you can also choose what Tableau shows when you are just hovering over the button by editing the tooltip text:

7. Once you are happy with your button, click OK and test the button:

Discussion

Navigation buttons allow you to navigate between your dashboards without applying any filtering.

When trying to test buttons in Tableau Desktop, you will need to press Alt (Option on Mac) and then click the button. When using Presentation Mode or on Tableau Server, you can select the button as normal without having to hold Alt or Option.

Another option for navigation between dashboards is using the Navigation Action; however, this requires using a sheet instead of a button, and therefore, could be quite confusing for the user.

Download Object

Although Tableau is an interactive tool, occasionally businesses need the ability to insert the dashboards into PDFs or PowerPoint documents. This Download object allows your end users to do just that. Remember, if you have important information within tooltips, anyone looking at the dashboard via a PDF or PowerPoint will have only a static version of the dashboard, making all tooltips redundant.

Problem

Your end user wants to download the dashboard as a PDF.

Solution

1. Drag and drop the Download object onto the canvas:

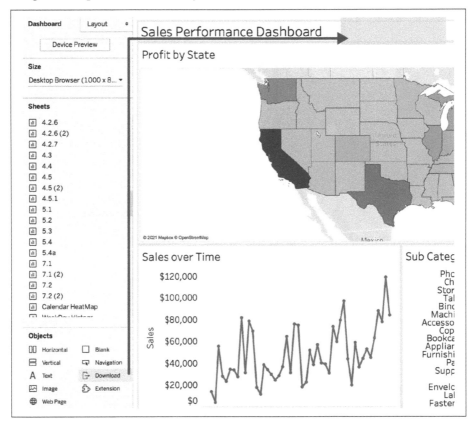

2. We can edit what the button is doing by selecting the drop-down by the object:

Similar options to the navigation button then pop up on your screen:

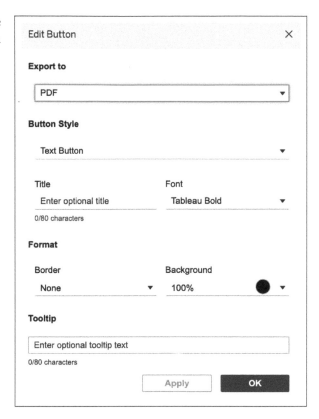

3. If you want to give the user the ability to download something other than a PDF, you can select other options:

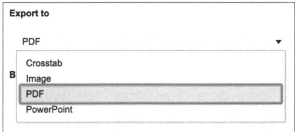

4. You have the same button type as the navigation, with either a custom image or text button.

5. You can also change the title, color, and tooltip of the button:

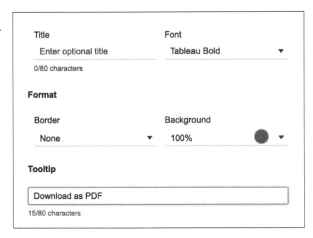

6. Test the button and see what exports:

Discussion

When you do have the "Download as PDF" button on a dashboard, clicking it opens a pop-up that asks what range you would like to include, the size, and some other options at the bottom:

Once you click OK, you are asked to save the PDF somewhere, and once you save, you have a static PDF of the dashboard.

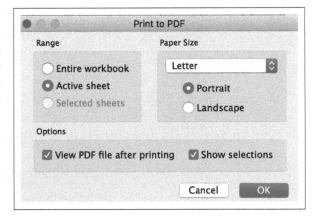

The Image, PowerPoint, or Crosstab options, allow you to save only the specific object type instead of the different options available for a PDF output.

Extensions

Dashboard extensions are web-based objects that enhance the functionality within your dashboard. They are created by Tableau and other members of the Tableau community. Extensions use the Tableau Extensions API. Tableau has a few categories of extensions, which include Advanced Analytics, New Viz Type, and Parameter:

Tableau includes a built-in gallery of extensions (*https://oreil.ly/NQf8s*).

Categories

- ⦿ All Categories
- ○ Advanced Analytics
- ○ Viz Formatting
- ○ Natural Language Generation
- ○ Monitoring & Stewardship
- ○ Custom Viz Actions
- ○ Write/Export
- ○ New Viz Type
- ○ Custom Filter
- ○ Parameter

Problem

You want to use the Image Map Filter extension, to allow an image to act as a filter on the dashboard.

Image Map Filter
by Tableau
☐ Sandboxed

Solution

1. Drag and drop the Extension object onto the dashboard:

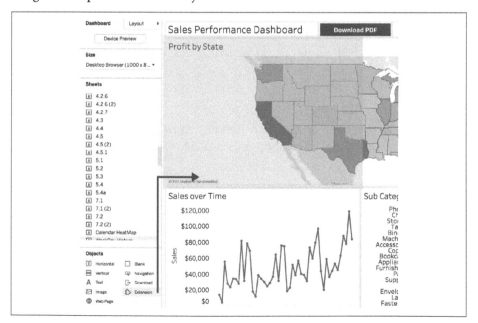

The built-in Extension Gallery pops up, allowing you to find your extension without the need to download it:

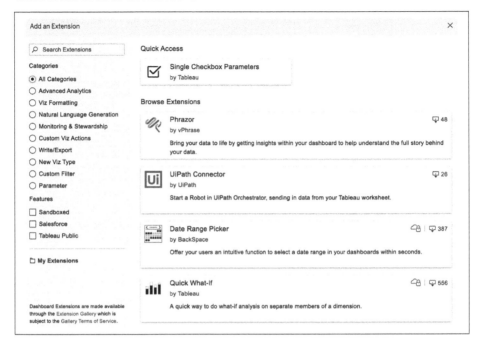

I recommend using extensions that are Sandboxed, as they run in a protected environment without additional access to the web, and therefore are more secure:

2. Find the Image Map Filter extension and select Add to Dashboard:

3. Once you have selected the extension, you will be asked to configure it. Which one you have chosen determines the type of configuration you will need to do. For this example, you have several options to choose from:

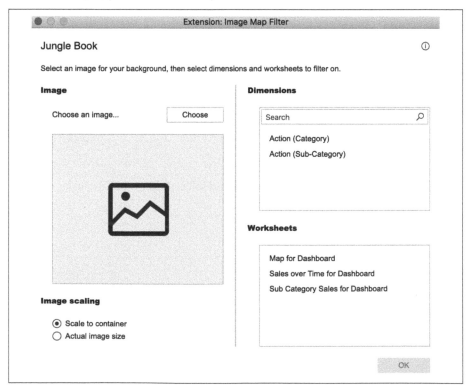

4. Select the image you want to use:

5. Select which dimensions to use for the filtering. This is based on actions on the dashboard. You also need to select which worksheets to apply it to:

6. Once we have our extension configured, we just need to tell Tableau that when a user clicks in a certain place to filter the dashboard to that category. To do this, click the pencil icon and the square at the end:

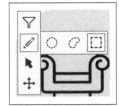

7. Now when you highlight a section of the image, it will give you a pop-up for the category:

8. Repeat for the other categories.

9. When you have finished and want to test the filter, you can click the filter icon at the top of the extension:

10. Now when you select an area, you will see the dashboard changes to that category selection. The image shows Furniture being selected:

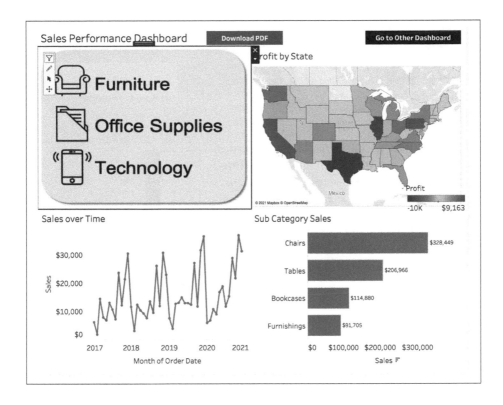

Discussion

Extensions add that next level of ability to your dashboards. They are built using an API and code.

Various types of extensions exist, so please read all the terms and conditions before applying and using an extension, especially if you have concerns around security.

To use extensions on Tableau Public, you need to check the box Tableau Public:

If you want to use extensions on Tableau Server, your Tableau Server Administrator will need to enable them.

19.3 Floating

Chapter 7 introduced horizontal and vertical containers. These allow design to be structured and aligned better. However, some designers prefer to float elements on a dashboard. Floating elements gives you more flexibility around the precise location and size of an element. You can make any element floating by clicking the drop-down and selecting Floating:

Or when you're bringing a new worksheet onto a dashboard, at the bottom of the Objects section you have the ability to select Tiled or Floating. If you select Floating, this will float the next element you add to the dashboard:

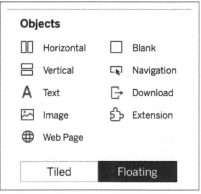

When designing dashboards, you can either float objects and sheets or use tiles and containers. When using tiled, the nest of containers can become never-ending in the Item hierarchy, which is why it is important to always start building any dashboard with a container —this stops the tiled action. One of the reasons people like to use float is because of the exact precision of an item; however, you can achieve the same effect when using containers if you use them correctly. I personally prefer to use containers and use floating for only specific things.

19.4 Toggle Containers On and Off Using Show/Hide

Using a show/hide button allows you to toggle a container on or off on the dashboard.

Problem

You want to maximize dashboard space but still give the user the ability to filter the dashboard.

Solution

1. Start with a dashboard that looks something like this:

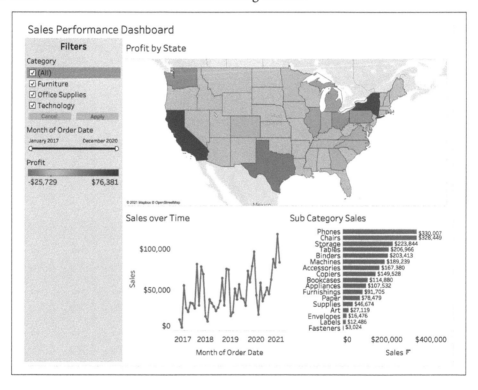

2. You want to give the user the ability to show or hide the filter panel.

To do this, we need to first select that container. Click the gray space below the color legend. You will know a container is selected by the blue outline:

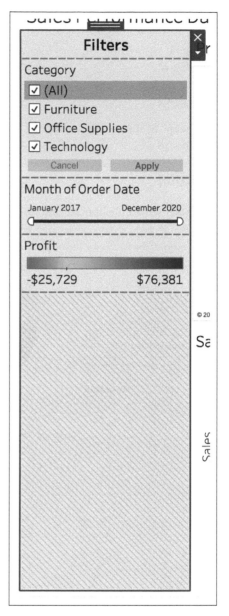

2. Click the drop-down arrow on the container and select Floating:

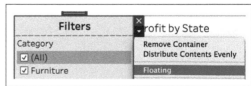

This will float the container and make the container smaller, because it resizes to fit the object:

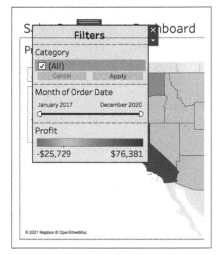

3. Make the container longer and place it in the position you would like:

4. To add the show/hide button, click the container drop-down arrow and select Add Show/Hide Button:

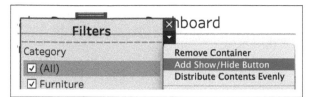

This currently adds a close icon to the dashboard:

5. To edit or change images or text, click the drop-down arrow and select Edit Button:

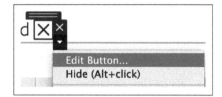

6. The options to customize this are similar to the navigation buttons covered in "Navigation Button" on page 624; however, this has the additional option of changing the appearance for when the container is shown or hidden:

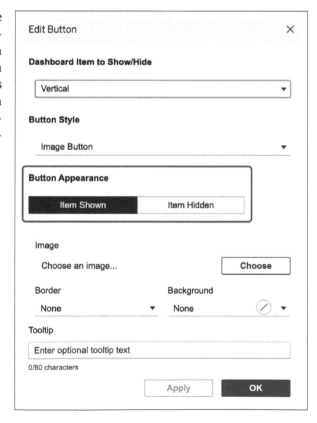

7. Now when you click the button (in desktop mode you need to press Alt or Option for it to work), the container will hide, and instead of the close icon we now have a hamburger icon:

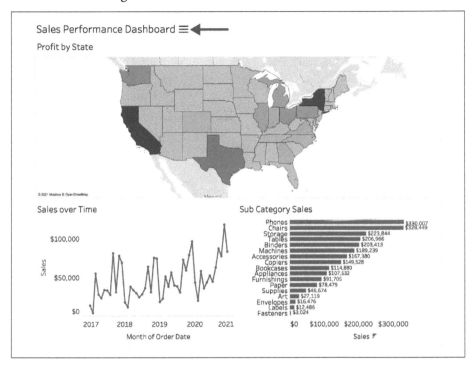

8. And if we select the hamburger icon, it will show the container again with all the filters inside:

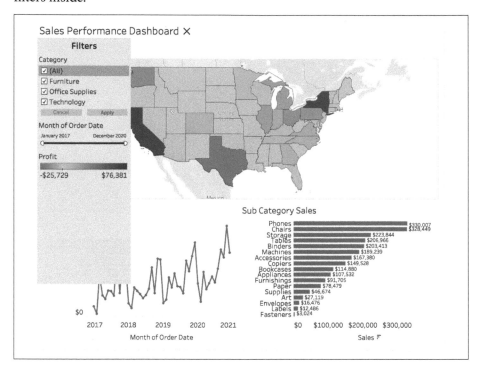

Discussion

Having the ability to show or hide containers, especially those with filters/parameters, allows the charts to have maximum space on the dashboard, rather than being confined with having a filter panel.

Top Tip 10: Show/Hide Containers

Tableau have also implemented the ability to use this Show/Hide Container button when you are using fixed container layout. Once you have clicked to highlight your container, click the drop down and select add Show/Hide Button:

This allows the container to collapse and expand within the outer container.

19.5 Device-Specific Dashboards

Once you have created your initial dashboard, you might have a request to make the dashboard fit onto a mobile or tablet device.

Problem

You want to add an automatic mobile dashboard for your viewers.

Solution

1. In a dashboard, on the left pane click the option to select Device Preview:

2. This will open a device preview at the top of the dashboard, where you select which layout you want to add, and it will show you the size of the layout on top of the dashboard:

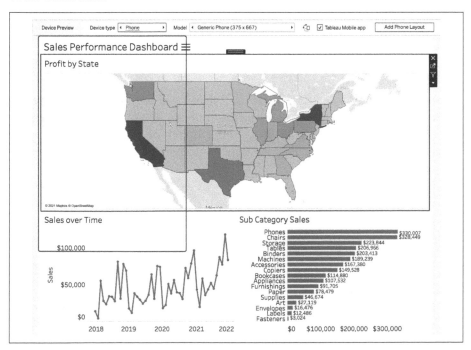

3. Once you've selected Add Phone Layout in the toolbar, Phone will now appear in the left panel with a lock icon:

This gives you an automatic phone layout, using the default layout as a baseline.

Discussion

When you add a mobile or tablet layout, Tableau will use the default layout to automatically place sheets, filters, and legends onto the device-specific dashboard. This follows a Z pattern, meaning anything in the top left of a dashboard will appear first on the mobile layout, followed by top right, bottom left, and finally, bottom right.

Tableau automatically creates a phone layout for every dashboard. If you go to Dashboard in the top menu, you will see two options: "Add Phone Layouts to Existing Dashboards" and "Add Phone Layouts to New Dashboards":

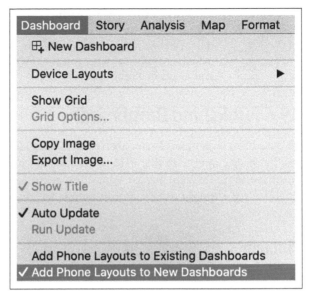

Although you have the automatic phone layout, if the layout is not how you want it, you can unlock the lock icon to allow changes to the dashboard:

If you unlock a device-specific layout and make changes to the default layout, these will not automatically apply to the device layouts. Change the device layouts only when you have finalized the default layout; otherwise, you could waste time changing the layout and miss something important.

19.6 Publishing Dashboards

So, you have your final dashboard ready to share with your stakeholders and managers, but how can you do that? There are several ways, depending on what license type you have with Tableau. You or your company might have Tableau Server or Tableau Online, or you might want to publish your dashboard to Tableau Public for the world to see.

Problem

You want to share your dashboards via Tableau Server/Online.

Solution

1. Now that you are happy with your dashboard, you want to publish it. To publish a dashboard, click Server in the menu at the top and select Publish Workbook:

2. This will ask you to sign into your Tableau Server or Tableau Online instance. Ask your IT or Tableau Administrator for more information.

Once you have signed in, Tableau will give you a few options:

3. You can choose which project to publish to and the name of the workbook; then you can add a description.

4. You can choose to add tags, and which sheets you would like to publish; for example, if you want to publish only the dashboards, you can select Only Dashboards:

5. Permissions to the workbook could be locked by your Tableau Administrator; please ask them for advice on permissions.

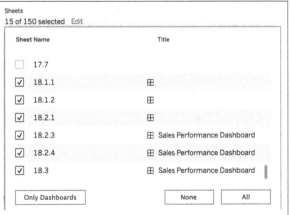

6. You also have the ability to show sheets as tabs, under More Options. This is useful when you have multiple dashboards without navigation buttons:

More Options
☑ Show sheets as tabs
☑ Show selections
☑ Include external files

7. Once you click Publish, the dashboard will be on your Tableau Server, and you have the ability to share with others who have the correct permissions.

Discussion

Publishing to Tableau Server or Tableau Online does require you or your business to have one or the other program. I recommend speaking to your Tableau Administrator before publishing any dashboards or data sources. Once you have logged in to Tableau Desktop, you will remain logged in to that server.

You have the ability to keep data live within the dashboard published; the data sources within these current dashboards are static Excel files, but if you have connected to a database, you have the option to schedule a refresh or to embed credentials to keep the data sources live and fresh.

The final option you have is to publish to Tableau Public, a public domain where everyone can publish and view dashboards. Do not publish any personal or sensitive data to Tableau Public, as anyone could download and use that information. Before you publish to Tableau Public, you will need to extract your data. Chapter 1 covers how to do this.

With Tableau Public being free, it is a great platform to showcase additional work (for example, charities and nonprofit organizations). It can also be used as a platform for organizations to communicate their data with the public and to embed the visualizations into a blog. Tableau Public should also be used to showcase your data visualization portfolio.

Once you have published your dashboard(s) to Tableau Server/Online/Public, Tableau Desktop will remember the last published location when you are connected; you will default to the same location next time you try to republish the workbook.

Summary

Dashboards are used to show multiple visualizations in one place. Using the techniques you learned in this chapter and Chapter 7, you should be proficient in creating dashboards and taking them to the next level. I recommend not making your dashboards too cluttered with a lot of charts and filters, but keep them simple with enough whitespace. It would be better to have a series of dashboards to allow the charts to breathe than to have all the charts squeezed onto one.

Part IV Conclusion

Part IV has given you the opportunity to develop your Tableau techniques that aid and enhance the analysis within Tableau. You have the ability to add additional interaction when using sets and parameters, including set actions and parameter actions. You have learned advanced Table Calculations and how to write and use LODs. Finally, you can take your maps to the next level with spatial calculations.

CHAPTER 20

My Top 20 Tips

Throughout this book, you have been given the opportunity to learn skills and techniques that are fundamental to understanding Tableau, and you should now feel comfortable in creating your own visualizations. To conclude this book, I want to share with you my Top 20 Tips to continue on your journey of learning and mastering Tableau Desktop. These are split into the categories of "Proactive" and "Technical."

20.1 Proactive Tips

Proactive tips are things that you can be doing outside of Tableau Desktop. All of these tips will help you along your journey with Tableau.

1. Create a Tableau Public Account

If you haven't already done so, I highly recommend creating a Tableau Public account. Tableau Public is a place for inspiration and sharing your own work. This can also become your portfolio to showcase your Tableau progression. Once you have started publishing to Tableau Public, never delete a visualization, because it shows progression of your Tableau skills as time goes on.

See Also

Tableau Public (*https://public.tableau.com/s*)

2. Interact with Community-Based Initiatives

Now that you have your Tableau Public account, it's time to start participating in Community-Based Initiatives and Projects. These initiatives are run by Tableau or Data community members. Makeover Monday is a weekly initiative that gives you a data set and a visualization to work with; you then have to create a better visualization using the data provided. This is a great way to practice your design and analytical skills. Workout Wednesday is another project I highly recommend, not just because I help organize it, but because it really helps you home in on specific Tableau skills (for example, workouts on table calculations or LODs). For these, you are given a data set, a visualization, and a set of requirements, and the aim of Workout Wednesday is to re-create exactly what you see by using the techniques specified.

A lot of other community initiatives are available to suit different requirements. Examples include Iron Quest, Preppin' Data, Project Health Viz, Sports Viz Sunday, Real World Fake Data, and Diversity in Data, to name a few. Visit Tableau's community hub for more information.

See Also

Makeover Monday (*http://makeovermonday.co.uk*)

Workout Wednesday (*http://workout-wednesday.com*)

Tableau Community Hub (*https://www.tableau.com/community*)

3. Follow Tableau Zen Masters and Tableau Ambassadors

I am a Tableau Zen Master and a Tableau Public Ambassador, but 42 other Zen Masters and 100 or more Tableau Ambassadors work in different areas of the software. Following these people on Twitter or LinkedIn will provide you with a lot of helpful tips and tricks, and hopefully provide you with a ton of inspiration:

- Tableau Zen Masters (*https://oreil.ly/yhq5u*)
- Tableau Ambassadors (*https://oreil.ly/Z1QY4*)

4. Invest in Additional Resource Books

You've successfully made it through this Tableau Desktop Cookbook, but plenty more books can help you on your Tableau Desktop journey. For example, Ryan Sleeper has two books, *Practical Tableau* and *Innovative Tableau*, both of which can take your Tableau Desktop skills to the next level. Preparing data can make your analysis a lot easier in Tableau Desktop. You can learn how in *Tableau Prep* by Carl Allchin.

Other data visualization books are also recommended, such as the following:

- Steve Wexler, Jeffrey Shaffer, and Andy Cotgreave, *The Big Book of Dashboards: Visualizing Your Data Using Real-World Business Scenarios* (Wiley, 2017)
- Andy Kirk, *Data Visualization: A Handbook for Data Driven Design* (Sage, 2016)
- Carl Allchin, *Communicating with Data* (O'Reilly, 2022, forthcoming)

5. Invest in Some Tableau Training

If books aren't the best learning resource for you, you can invest in some Tableau training. Lots of free tutorials are also available through community blogs or YouTube videos. Tableau also has an e-learning platform that you can take advantage of.

See Also

The Information Lab, YouTube (*http://til.bi/youtube*)

Andy Kriebel, YouTube (*https://www.youtube.com/user/kriebela*)

Tim Ngwena, YouTube (*https://www.youtube.com/user/tfngwena*)

Tableau E-Learning (*https://www.tableau.com/learn/training/elearning*)

6. Create Your Own Blog

I started writing a blog while I was in training. The reason I started a blog was to help solidify my learning, but also to refer back to in the future. Other people can also find these blog posts useful to learn from, and seeing an explanation in someone else's terms frequently helps the audience fully understand a technique.

7. Keep on Top of New Features and Get Involved in the Beta Program

Tableau aims to release a new version of Tableau every quarter, which is normally packed with new features. Tableau has a beta version, which allows anyone to go and investigate some of the new features prior to release. This is a great way to get on top of the new features, before you have access to that at work, and you also have the ability to provide product feedback.

See Also

Tableau New Features (*https://www.tableau.com/products/new-features*).

8. Join the Tableau Community Forums

The Tableau Community Forums is a place for anyone to ask Tableau-based questions. Forums could span all of Tableau's products. You can also help answer some of the questions on the site. The community forums are also the place to suggest new product ideas to Tableau or spend some time upvoting other people's ideas.

See Also

Tableau Community Forums (*https://community.tableau.com/s*)

9. Find Your Local Tableau User Group

Tableau User Groups are run by the community, for the community. They are a great place to learn about feature updates, different use cases for Tableau, and tips and tricks on how to use the product. These meetups can take place in person, or plenty of user groups take place virtually.

See Also

Tableau User Groups (*https://usergroups.tableau.com*)

10. Attend a Tableau Conference

Tableau Conference happens every year, across different parts of the world. Attending Tableau Conferences allows you to stay up-to-date with the latest features and product developments. There you can attend training courses run by Tableau, but most importantly, you can connect with thousands of like-minded people who all have a passion for Tableau. There really is nothing like a Tableau Conference.

Here is a selection of pictures from Tableau Conference 2019:

20.2 Technical Tips

Throughout this book, you might have seen Top Tips. These are my Top 10 Technical Tips. The selection of tips are techniques that I have found incredibly useful when building anything within Tableau; they add that additional context or look-and-feel, and they can also be used to save time.

Here is a summary of where you can find them in the book:

- "Top Tip 1: Adding Totals from the Analytics Pane" on page 156
- "Top Tip 2: Custom Color Palettes" on page 159
- "Top Tip 3: Follow Tableau Zen Masters and Tableau Ambassadors" on page 277
- "Top Tip 4: AVG(0)/Dummy Axis" on page 434

And finally, once you have practiced most of the techniques in this book, you should get yourself certified. Tableau offers three Tableau Desktop–specific exams: Tableau Specialist, Tableau Certified Associate, and Tableau Certified Professional. This book has given you the majority of the skills needed for becoming a Tableau Specialist and Tableau Certified Associate. To become a Tableau Certified Professional, you will need more practice on some of the harder desktop skills and practicing to develop good analytics with Tableau.

Thank you for taking the time to read this book; I hope you enjoyed it and have learned something new. You should now feel confident building basic and complex data visualizations in Tableau Desktop, and you can use the recipes throughout this book to help improve your analytical skills. Moving forward, you should now be able to apply the skills learned while using data visualization best practices, for creating interactive dashboards to support business questions.

All the best for your Tableau journey,
—Lorna Brown

Index

Boolean data type, 38
borders on sheets, 297
box plots, 420-424
buffer calculation, 604-606
bullet charts, 137-141
buttons
    navigation button, 624-626
    show/hide, 637-642

C

calculated fields, 54-59
    Basic, 55
    calculation window, 113
    color, 99
    Level of Detail, 55
    parameters, 507
    row-level calculations, 55
    Table Calculations, 55
calculations
    date parameter, 516
    dimension switcher, 524
    drill downs, 539
    LODs (level of detail), 567-569
    parameters, 512
    rank calculation, 164
calendar heatmaps, 317-323
    sizing, 331
Cards, 46, 59
Cell average line, 102
charts, 44
    (see also specific charts)
    area charts, 351
    bullet charts, 137-141
    control charts, 561-566
    donut charts, 433-439
    dot plot, 408-413
    histograms, 142-145
    pie charts, 430-433
Circle mark type, 383
clusters, scatter plots, 390-394
color
    backgrounds, 297
    BANs, 168
    color palette
        automatic, 159
        types, 159
    color-blind users, 99
    density color palette, 390
    Density marks, 589

highlight tables, 159, 159
Intensity slider, 390
parameterized reference lines, 98
segments, 63
stacked bar charts, 118
color legend, 263, 268
    background colors, 297
    hiding/showing, 66
    padding, 301
Color Marks card
    conditional grouping, 112
    visual grouping, 106-107
Columns, 59-61
combined sets, 488-493
command buttons, 340
Community-Based Initiatives and Projects, 652
Condition filter, 72
conditional categories, 112
conditional grouping, 111-114
conditional sets, 482-486
conditional statements, calculation window,
    113
Connect pane, 3, 18
    Search for Data option, 4-5
connected scatter plots, 401-407
constant line (bar charts), 94-95
constant set, 493-496
containers
    objects, 277-285
    toggling, 637-642
continuous dates, 191
continuous line charts, 196-198
control charts, 561-566
correlation, Window Calculations, 561
CSV (comma-separated values), connections,
    10

D

Dashboard Actions pop-up box, 270
dashboards, 44
    actions, 268, 270
        filter actions, 273
        highlight actions, 611-616
        Hover option, 274
        Target Sheets, 272
        URL actions, 616-621
    backgrounds, 297-301
    BANs, 165-169
    building, 259-268

E

Edit Reference Line option, 96
Edit Table Calculation option, 162, 208
    options, 209
edit, calculation, 57
    calculated fields, 521
    Edit Table Calculation option, 162
    forecasting, 203, 211
    LODs (level of detail), 582
    LOOKUP functions, 212
    maximum value, 582
    offset, 557
    percent differences, 215
    reference lines, 96
    sets, 495
    specific dimensions, 208
ELSE statement, 113
Excel, 5
Exclude LOD, 573-577
Explain Data, 399-401
extensions, dashboards, 630-636
extracts, 38-40

F

fields
    ad hoc, 149
    calculated, 54-59
    Columns, 60
    hiding/unhiding, 111
    hierarchy, 100
    highlight actions, 615
    Rows, 60
files, connecting to, 5
filled maps, 245-248
filters
    dashboards, 268-276
    difference calculation, 174
    order of operations, 572
    Top N, 114-116
Filters shelf, 67-72
    Category field, 68
    Condition filter, 72
    continuous filters, 72-76
    Top filter, 72
    wildcards, 72
Fixed LODs, 577-581
fixed sets, 493-496
float/integer parameter, 505-510
floating elements, dashboards, 636

font formatting, 80-85
forecasting, line charts, 202-203
Format pane, 84
    Alignment, 84
    Borders, 84
    Font, 84
    Lines, 84
    Shade, 84
Format Workbook menu, 81
formatting
    date parameter and, 517
    dates, 204-206
    display format, 506
    fonts, 80-85
    Sheet level, 84
    story points, 462
    text, 177-179
    tooltips, 335-341
Full Outer joins, 34

G

Gantt charts, 441
    barcode charts, 445-448
    waterfall charts, 448-455
general sets, 480-482
grouping
    conditional, 111-114
    discrete, 107-111
    manual, 103-105
    visual, 106-107

H

heatmaps
    calendar heatmap, 317-323
    marginal bar charts, 323-333
highlight actions, dashboards, 611-616
highlight tables, 156-160
    color palettes, 159
histograms, 142-145
horizontal bar chart, 89-94
    converting to vertical, 91
    stacked, 125
horizontal containers, 277
Hub and Spoke map, 593
    Hub start and endpoints, 596

I

IF statement, 113

About the Author

Lorna Brown is a Tableau Zen Master and Tableau Public Ambassador. She started her Tableau journey as part of the world-leading analytic training program called the Data School. Her journey continued by working in various industries, using Tableau to help companies see and understand their data. Lorna is also the coleader of two Tableau community initiatives. Workout Wednesday, a weekly challenge to re-create data-driven challenges, and Tableau Tip Tuesday, where she provides video tutorials on the tips and tricks she has learned in Tableau. Lorna loves to give back to the Tableau community and helps host the North West UK Tableau User Group.

Colophon

The animal on the cover of *Tableau Desktop Cookbook* is the Eurasian wolf, *Canis lupus lupus*. This subspecies of the gray wolf (*Canis lupus*) originally ranged widely, across Russia and China in the east to India and Saudia Arabia in the south and across all of Europe. They now occupy only disparate small areas of that range.

The gray wolf is the largest species of canids, with males measuring an average of 4 feet long and 4 feet tall, with an average weight of 120 pounds. Females are slightly smaller. The color of their thick coats varies among individuals, with wolves ranging from black to white sometimes mingled with shades of brown.

Wolves are intensely social creatures with a complex family life. Gray wolves form lifelong pair bonds and breed in spring, giving birth to about six to seven pups. Wolf packs form around a central pair, occupying and hunting in a home range of many square miles. Pups are raised together by the breeding pair and the pack. Young wolves stay with the pack they were born into for one to three years.

Wolves eat a variety of foods, sometimes hunting large herbivores such as deer. They also hunt opportunistically, and at times will kill smaller prey such as wild boar, hares, and livestock animals.

Gray wolves figure regularly in legends and iconography across their range, including European folktales such as "Little Red Riding Hood" and the legend of the wolf who nursed Romulus and Remus, the mythic founders of Rome; Asian stories such as the Chinese folktale "The Wolf of Zhongshan"; and the tales of the *Panchatantra*. Many of the animals on O'Reilly covers are endangered; all of them are important to the world.

The color cover illustration is by Karen Montgomery, based on a black and white engraving from J. G. Wood's *Animate Creation* (1885). The cover fonts are Gilroy Semibold and Guardian Sans. The text font is Adobe Minion Pro; the heading font is Adobe Myriad Condensed; and the code font is Dalton Maag's Ubuntu Mono.